In Search of Human Evolution

In Search of Human Evolution

Field Research in Diverse Environments

MICHAEL H. CRAWFORD

OXFORD
UNIVERSITY PRESS

Oxford University Press is a department of the University of Oxford. It furthers the University's objective of excellence in research, scholarship, and education by publishing worldwide. Oxford is a registered trade mark of Oxford University Press in the UK and in certain other countries.

Published in the United States of America by Oxford University Press
198 Madison Avenue, New York, NY 10016, United States of America.

Library of Congress Cataloging-in-Publication Data
Names: Crawford, Michael H., 1939– author.
Title: In search of human evolution : [field research in diverse environments] / Michael H. Crawford.
Description: First edition. | New York : Oxford University Press, [2024] | Includes bibliographical references and index.
Identifiers: LCCN 2023054370 (print) | LCCN 2023054371 (ebook) | ISBN 9780197679401 (hardback) | ISBN 9780197679418 (epub)
Subjects: LCSH: Human evolution—Fieldwork.
Classification: LCC GN281 .C735 2024 (print) | LCC GN281 (ebook) | DDC 599.93/8—dc23/eng/20240131
LC record available at https://lccn.loc.gov/2023054370
LC ebook record available at https://lccn.loc.gov/2023054371

DOI: 10.1093/9780197679432.001.0001

Printed by Integrated Books International, United States of America

Contents

Preface

Fieldwork in different geographical regions of the world with unique social organizations and environmental conditions provides comparative information on evolutionary questions. Currently, it is extremely difficult to conduct field investigations in most parts of the world due to a series of factors, such as infectious disease epidemics, COVID-19, political upheaval, wars, illegal drug smuggling, and past mistreatments of indigenous populations. This volume examines (1) the nature of a number of evolutionary questions; (2) sources and germination of the experimental design and its limits; (3) methods for obtaining permissions to conduct field investigations on an international, national, or local level; (4) methods employed for data collection; and (5) data analyses and interpretation.

Contemporary anthropological geneticists utilizing DNA analyses often receive specimens from field researchers and collaborators and then decide what questions are possible or appropriate. There are increasing numbers of researchers who address anthropological issues without ever having conducted actual fieldwork. Instead of working with specific communities, they work on data. Instead of collecting appropriate data needed to address specific questions, they seem to get what data they can find and then determine what questions they might attempt to answer. They rarely design a field project, collect data, and then test specific hypotheses. Often the exact sources of the samples sent to them are in question. Who was sampled and where? A number of studies have published conflicting genetic results from supposedly the same populations. These differing genetic results reflect population structure variation and geography. This book demonstrates the need to undertake fieldwork and engage with the communities of interest and in partnership with the researchers.

Given more than 50 years of field research, I would like to acknowledge a large number of willing participants from an assortment of universities and research institutes. Each researcher who took active part in my field investigations is listed within the 12 chapters of this volume together with the role they played in each project.

I would like to acknowledge the contributions of colleagues in the preparation of this volume. Dr. M. J. Mosher, who originally participated in the organization and completion of the Mennonite research program, checked the prose and accuracy of the data discussed in the Mennonite chapter. She also edited text in this volume.

My maternal relative (third cousin) from Blagoveschensk, Siberia, Nadezhda Jakovleva, and her daughter Yana Mescheryakova helped reconstruct my genealogy using Russian records and documents and checked the accuracy of the verbally transmitted family history.

I would like to thank my son, Kenneth Crawford, for preparing illustrations, editing text for this volume, and struggling over the style of the book. My wife, Carolynn, patiently tolerated my travel to many exotic regions of the world and provided emotional support for my academic career.

1
Introduction to Fieldwork and Evolution

This volume focuses on the application of field investigations (i.e., research conducted in places other than one's own laboratory) to the measurement of human evolution in unique and diverse populations, particularly those groups that experienced migration and are located in contrasting environments of the world. A chapter entitled "The Importance of Field Research in Anthropological Genetics: Methods, Experiences and Results," published by Cambridge University Press in a volume *Anthropological Genetics: Theory, Methods and Applications*, introduces readers to the impact of field investigations to the understanding of human evolution (Crawford, 2007d).

Why would researchers be willing to subject themselves to scorching heat, swarms of Anopheles mosquitoes, and/or frigid conditions in the mornings or nights as they struggle to use a smelly outhouse? Field research helps us answer a number of basic and universal questions: Who are we? Where did we come from? How did we get here?

Field research (particularly comparative investigations in other communities and countries) is an essential tool of anthropological genetics—a synthetic discipline that applies the methods and theories of genetics to evolutionary questions posed by biological anthropologists (Roberts, 1965, 1968; Crawford, 2007d). It provides (1) a comparative dimension to a study; (2) a possible time dimension; and (3) unique demographic events (such as migration) and social structures of populations. Fieldwork provides diverse environmental and cultural conditions that are not possible in laboratory situations for testing specific hypotheses (Crawford, 2007a). Traditional human genetic studies focus on random samples of communities or hospital patients, and these conditions are traced through specific extended families. In contrast, anthropological geneticists often conduct research away from their laboratories in the field under more challenging social and environmental conditions. For example, the Black Caribs (also known as Garifuna) discussed in Chapter 5 of this volume, offer anthropological geneticists an opportunity to examine unique social structure and apply statistical

In Search of Human Evolution. Michael H. Crawford, Oxford University Press. © Oxford University Press 2024.
DOI: 10.1093/9780197679432.003.0001

methodology such as the half-sib method for analyzing the variances associated with quantitative traits (Crawford, 1984). This social organization is possible because of the fluidity in the extended families of the Garifuna. According to cultural anthropologist Nancie Gonzalez (1984), the strongest and most enduring social unit among the Garifuna is the sibling group consisting of one mother but with a number of different fathers. Males periodically join the families and father children and then migrate for work to different regions of Central America. Thus, Garifuna families typically consist of a number of half sibs, who share a common mother's genome but with an assortment of different fathers and Y-chromosome markers.

History provides a time dimension to human gene pools that have undergone fission and/or transplantation to vastly different environments. For example, the indigenous Tlaxcaltecans of Central Mexico offer a unique opportunity to add a time dimension of more than 300 years to the study following the fission of the Valley populations into groups transplanted by the Spanish Crown to different regions of Mexico. Chapter 3 of this volume examines the genetic results of transplantation in 1525 of families and garrisons from the Valley of Tlaxcala to Cuanalan (a municipio of Acolman in the adjoining Valley of Mexico) and relocation in 1570 of 400 families from Tlaxcala to Saltillo with its desert-like environment of northern Mexico.

Background and Experiences

Who in their right mind would willingly subject themselves to the rigors of fieldwork under severe conditions of heat, cold, or voracious insects? Chapter 2 introduces the background and experiences that prepared me for fieldwork in diverse regions of the world, such as Mexico, Guatemala, Newfoundland, Belize, St. Vincent Island, Dominica, St. Lawrence Island, Ireland, Wales, Alaska, Evenkiya, and populations of Kamchatka, Siberia, the Aleutian Islands, Tiszahat, Hungary, and Sukhumi, Abkhasia. My life experiences inadvertently prepared me for extreme travel, existence, and research under challenging field conditions. Childhood experiences that prepared me for future fieldwork included (1) residence in Shanghai, China, during Japanese wartime occupation. My family experienced shortages of food, absence of medical services, and scarcity of housing; (2) travel from China to a displaced person's camp in Tubabao, Philippine Islands, with survival for nine months living in a military surplus canvas tent, while existing under tropical conditions including heat, humidity, and periodic typhoons. This camp was located on a small island off the southern coast of Samar, Philippine Islands; (3) relocation by boat and train from the Philippine Islands to a displaced persons camp in Uranquinty, New South Wales, Australia. My family survived internment in the Philippine Islands despite no running water or electricity but with clouds of voracious mosquitoes. The morbidity and mortality of the displaced persons living under such conditions was exceptionally high from a variety of infectious and chronic diseases. After one and one half years, the survivors of the Tubabao displaced persons camp were scattered throughout the world, relocated to Brazil, Australia, United States, Ecuador, and Argentina (Nash, 2002). My family was transplanted from the Philippine Islands to a displaced persons camp in Australia. All of these life events occurred before settling in the United States at the tender age of 13 years.

In this volume, I review how multidisciplinary field investigations on human evolution are organized, specify the methods for selection of accompanying specialists, and explain how the indigenous communities are informed about the risks and benefits of the research. Because of past ethical abuses of indigenous peoples, currently, it is more difficult to conduct field research in some regions of the world. Often indigenous people are suspicious of the "gringos" (a derogatory term in Spanish and/or Portuguese for a foreigner, usually an English speaker of European descent) who asks embarrassing questions and wants samples of their body fluids, such as saliva or blood. Some governments, such as Russia, currently prohibit all exportation of DNA from indigenous populations.

Anthropological genetic research is a partnership with both the scientists of that country and the specific communities. The research teams should include local researchers and students so that in the future they can conduct their own research without outsider interference. Feedback to the community, results and discoveries, must be shared with the study populations, local universities, and medical institutions. Upon the completion of the Aleutian Island project, more than 100 Aleuts attended my lectures in Anchorage when I provided feedback on questions of interest to the community. They were particularly interested in the Aleut origins: Who are we? Where did we come from? How did we get here? A summary of the results of research on the Aleutian Islands was published in a community newsletter *The Aleutian Current* and was available to all Aleuts (Crawford, 2006). Individual requests for genetic (DNA) results were provided; however, this confidential feedback was limited only to individual participants of the study.

What's Evolution?

In this volume, *In Search of Human Evolution*, I define evolution as a change in the frequencies of genes in a population (gene pool) over time. Four forces of evolution affecting the frequencies of genes in a population include (1) mutations; (2) natural selection; (3) gene flow; and (4) genetic drift.

Mutations are spontaneous genetic changes in the genome during replication, and they are a source of new genes entering a gene pool. On a species level, mutations are the only source of new genetic material into the gene pool.

Consequences of mutations are as follows:

1. Origins of rare genes can be traced to specific mutant individuals in small populations. For example, Amish populations in Pennsylvania and Ohio with few founders have a number of deleterious mutations, such as hemophilia B, pyruvate kinase deficiency, and limb girdle muscular dystrophy.
2. Source of variation that selection can affect. Best examples involve malaria operating on an assortment of mutations in different regions of the world. Glucose-6-Phosphate-Dehydrogenase (G-6-PD) deficiency, Duffy null, and Hemoglobin S and C are examples of a variety of mutations that affect the lifecycle of Plasmodium vivax or Plasmodium falciparum and provide resistance to malaria.

Natural selection is a force of evolution characterized directly through the calculation of fitness (w) and selection coefficient (s) in a population, or indirectly through the examination of the resulting genomic variation. The survivor with the most children has the highest fitness and his/her genes become most numerous in the subsequent generations. Selection on a molecular level can be detected by comparing variation in specific regions of the genome with surrounding genomic regions.

Consequences of natural selection include the following:

1. Increases in the frequencies of favorable genes in a population—that is, causing evolutionary change
2. Balanced polymorphism where heterozygotes have an advantage over both homozygotes
3. Selection can be directional (e.g., lactase enzyme and ability to drink milk) or balancing selection (smallpox, plague, and malaria)

Gene flow is the process through which genes are introduced from one gene pool of a breeding population into the gene pool of another population. This movement of genes is usually associated with migration. For example, the creation of the Mestizo gene pools in Central America was due to Spanish and African gene flow associated with cultural contact and conquest (Munoz and Crawford, 2021).

Consequences of gene flow are as follows:

1. Introduces new genes into another environment
2. Increases heterozygosity and decreases homozygosity
3. Makes populations more similar genetically

Genetic drift is the cumulative effect of all random processes modifying the frequencies of genes in small populations (genetic isolates).

> Isolated populations are currently defined as subpopulations deriving from a relatively small number of individuals (founders) who became isolated from their ancestral group (e.g. through the settlement of a new territory) and/or had experienced a significant reduction in population size. (Lopes et al., 2016, p. 110)

Causes of genetic drift include (1) random variation that occurs from the passing of gametes from one generation to the next; and (2) unique historical events, such as founder effect or genetic bottleneck.

The consequences of genetic drift on populations include the following:

1. Increase in homozygosity at the expense of heterozygosity
2. The genetic differentiation of small populations
3. Reduction in variation within subpopulations

Contents of This Book

This volume, *In Search of Human Evolution*, reviews my search of more than 50 years for evidence of human evolution under a variety of environments and driven by unique historical events, such as population transplantation. This chapter introduces the reader to field research and the current evidence of human evolution. Chapter 2 describes those events in my life history that prepared me for field investigations and academic wars. Each subsequent chapter outlines the nature of the evolutionary question, the source and germination of the idea for field investigation, methods of data collection, analyses, and conclusion. Chapter 3 examines the effects of Tlaxcaltecan (Mexican) population fission, transplantation followed by massive gene flow from Spanish invaders (Crawford, 1976). The primary question posed by the Tlaxcaltecan project is this: Do populations evolve when separated geographically over time? Chapter 4 discusses the origins of Irish Travelers (also known as Tinkers or Itinerants), their social isolation, and genetic differentiation from the surrounding Irish populations through genetic drift. Chapter 5 focuses on the Black Caribs (Garifuna) of Central America and the Caribbean and documents an evolutionary success story, their forced migration from St. Vincent Island, their numerical expansion, and colonization of most of the Atlantic coast of Central America. Their evolutionary success resulted from their culture and extractive efficiency (a combination of fishing and horticulture) and genetic adaptation through mutations that provided resistance to malaria. Chapter 6 examines the genetic consequences of population fission of a religious isolate, the Anabaptist Mennonites of Kansas and Nebraska, their genetic differentiation, and because of excellent family records, their histories, and the genetics of biological aging. Chapter 7 considers the genetic structure of Siberian indigenous populations and their role in the peopling of the Americas. Chapter 8 documents the genetics of the Aleut (who refer to themselves as Unangan) expansion from Siberia into the 1,600-mile Alaskan Archipelago and considers the evolutionary consequences of population fission, gene flow associated with Russian colonization, and their genetic structure as assessed through molecular genetics. Chapters 9 and 10 summarize a number of more limited research investigations conducted by the Laboratory of Biological Anthropology, some successful and others terminated for a variety of reasons. Chapter 9 reviews the application of bio-demography to two agricultural human populations, residing in Italy and Hungary: Valle Maira and Tiszahat. The first example of the application of bio-demographic methodologies focuses on the breakdown of reproductive isolation in Valle Maira, an Italian mountain valley located south of the city of Torino. Church record analyses from Tiszahat, Hungary, permit the examination of the genetic consequences resulting from a shift in the international border between Russia and Hungary, which divided the Hungarian villages of Tiszahat. Chapter 10 applies genetic epidemiological methodology (genetic–environmental interactions) in the study of a complex infectious disease, lymphoma in a baboon (*Papio hamadrayas*) colony of Sukhumi, Abkhasia. Chapter 11 examines the molecular evidence for the origins and ancestry of the Basque populations of western, mountainous Spain and tests the likelihood of three hypotheses generated by linguistic and standard blood genetic markers. The concluding chapter (Chapter 12) summarizes the scientific importance of field investigations to our understanding of human evolution and considers the ethics of such programs.

Appendix A features a copy of a letter of acceptance written by Charles Darwin for his nomination for membership into the Argentine Academy of Sciences—an honor that I shared with him, one century later. Appendix B includes a list of Ph.D.s in biological anthropology from the Laboratory of Biological Anthropology (LBA) at the University of Kansas, who were involved in the research discussed by this volume.

2
Background and Preparation for Fieldwork

Genealogy and Migration

I was born on July 25, 1939, in Shanghai, China, of Russian/Scottish American parents, namely Tamara Innokentievna Lovtsova and Herman Charles Crawford. Following Russian tradition, I was not given a middle name but assumed my father's first name, Herman or German, since the Russian (Cyrillic) alphabet lacks the letter "H" but uses "G" in its stead. Thus, my father's Russian name is "German," after the lead character and gambler of the Russian opera *Pikovaya Dama* (*Pique Dame*) composed by P. I. Tchaikovsky and his librettist brother, Modest. Therefore, my legal Russian name in China was Mikhail Germanovich Crawford. However, professionally I went by the name Michael Herman Crawford, or M. H. Crawford.

Table 2.1 summarizes the migration patterns of my family during the 18th to 20th centuries on both the maternal and paternal lines. My mother's grandfather, Vsevelod Nabokov, was born in Pskov Province. He first served in the Russian military, trained in gold-mining engineering at the highly acclaimed Mountain Institute of St. Petersburg, and worked as a mining engineer at the Altai mines. Nabokov was invited by the governor of the Amur region of Siberia to establish gold-mining operations in Blagoveschensk. The Russian Revolution drove my mother's family first to Harbin and later to Shanghai. My paternal grandfather born in Dixon, Illinois, was also a gold-mining engineer, trained in South Dakota. He managed the tsar's gold mines in Chita. My family was forced by the Bolshevik Revolution in Russia to escape from Siberia to Shanghai. The Chinese communist takeover of China resulted in a rapid departure to the Philippine Islands. Because my family lost its US citizenship, making us stateless, we could only move to United Nations displaced persons (DP) camps in the Philippine Islands and Australia. This constant family relocation does select for personalities who are able to adapt to a wide variety of environmental and political situations.

In Search of Human Evolution. Michael H. Crawford, Oxford University Press. © Oxford University Press 2024.
DOI: 10.1093/9780197679432.003.0002

Table 2.1 Schematic Representation of Family Migration Summarized for the Maternal (Nabokov) and the Paternal (Crawford) Lines

Maternal Family: St. Petersburg, Russia → Blagoveshensk, Siberia → Harbin, Manchuria → Shanghai, China → Tubabao, Philippines → Uranquinty, Australia → Sydney, Australia → Seattle, Washington
Paternal Family: Dixon, Illinois → Chita, Siberia → Harbin, Manchuria → Shanghai, China → Tubabao, Philippines → Uranquinty and Sydney, Australia → Seattle, Washington

Shanghai, China

My father, Herman Charles Crawford, was born in Chita, Siberia, son of Charles Hamilton Crawford, and Siberian mother, Nina Rogova. In 1927–1930, my father moved for educational and work opportunities from Chita to Harbin, Manchuria, and then to Shanghai, where he received formal training as an electrical engineer and installed Carrier air conditioning systems into the Grand and Cathay Theaters, the first air-conditioned cinemas in Asia. Because of the extreme summer heat in Shanghai and the absence of home air conditioning units, Chinese oligarchs and international visitors to Shanghai, China, spent considerable time watching movies while cooling off in the air-conditioned theaters. Herman Crawford hosted the Dai Li Lama when the Tibetan religious leader visited Shanghai. He also served as the manager of the famous Russian basso Feodor Ivanovich Chaliapin during Chaliapin's tour of the Far East. My father eventually managed four cinema theaters (Nanking, Cathay, Grand, and Capitol) in Shanghai while employing more than 2,000 Chinese workers in these theaters. Without any formal linguistic training, he mastered three Chinese dialects (Shanghai, Mandarin, and Cantonese) and spoke each of them fluently and without an accent. He learned both Russian and English as a child and had complete fluency in both languages.

My mother, Tamara Inakentivna Lovstova, attended primary school in Blagoveschensk (a Siberian city on the Amur River) but completed high school with honors at the Sacred Heart Academy in Shanghai, China. The family (consisting of Sofia Lovstova, Tamara, and her two surviving brothers) first moved to Harbin, Manchuria, to avoid the advancing Bolshevik army that had executed her White Russian father. The third brother of Tamara Crawford had died from upper respiratory illness. Harbin (Manchu word meaning a place for drying fishnets) was a small rural settlement on the Songhua River.

Harbin had become the largest enclave of Russian emigres. In 1931, following a false flag Mukden incident, staged by the Japanese military, the Japanese blew up a portion of their own railroad near the town of Mukden and blamed the Chinese nationalists. This incident provided a pretext for the invasion of Manchuria and the creation of the puppet state of Manchukuo. Due to this invasion, the majority of Russians left Harbin and fled to Shanghai, China. The USSR sold the Chinese Eastern Railway to the Japanese in 1935, precipitating the further migration of the Russians out of Harbin. Although my grandmother Sofia Lovstova remained in Harbin, until her death from breast cancer, my mother relocated to Shanghai, where she lived with her aunt and completed her high school education. On completion of high school with honors, she was hired by one of the theaters managed by my father. They were married in 1934 in Shanghai. My sister, Marion, was born in September 29, 1935, followed by my birth four years later.

Japanese Occupation of Shanghai

I was born in Shanghai, China, just prior to the Japanese military occupation of the city. In December 7, 1941, following the attack on Pearl Harbor, the Japanese Imperial Army, which had earlier encircled the city in 1937, invaded the Shanghai International Settlement.

Before the Japanese invasion, my family resided in a luxurious two-story penthouse on the roof garden of the Cathay Theater that was located across the street from the neoclassical Cercie Sportif Francais (the French Club) with its striking red-clay tennis courts. I used to watch tennis matches from the roof of the Cathay Theater. After the Japanese invasion, my family was evicted from the penthouse by the Japanese military, who had installed antiaircraft guns on the flat roof of Cathay Theater. During World War II, both Japanese troops and their horses were housed in the theater. Initially, we hoped to move into a large apartment owned by a German relative, by marriage, Rolf von Heinkel (after the war he changed his surname to Rolf Hennequel) and my mother's aunt, Claudia (Clasha). However, being an academic, with a Ph.D. (30 years earlier) from my alma mater, the University of Washington, he preferred privacy, and Rolf refused to accept such a move by my family. We were fortunate to be able to move into a house owned by my father's aunt, Dariya (better known as Dasha), and her veterinarian husband, Boris. He cared for the racehorses at the Shanghai Race Club track, which reopened soon after Japanese occupation and continued to host races throughout the war. With all four of the movie theaters in Shanghai eventually closed during the Japanese occupation, my father lost his position as manager but was able to eke out an existence by betting on horse racing at the Shanghai track. These money-making bets on the horse races were made possible because Boris provided detailed information on the condition of each horse and the projected winners of most races. During the Japanese occupation, proximal to the house where we lived, was a prisoner-of-war camp. Such a camp was graphically depicted by J. G. Ballard in his award-winning novel and movie *Empire of the Sun*. Ballard survived a prisoner-of-war camp while being interned from 1942 to 1945 at the Lung Hua Center of Shanghai (Ballard, 1985).

My family was fortunate in not being interned in a formal prisoner-of-war camp because my father's American citizenship had been revoked by the US Embassy following the death of his father but preceding the Japanese occupation. It turned out that my grandfather had committed bigamy by not divorcing his first wife, Bertha Todd. As a result, my family became stateless—that is, people without a country. The city of Shanghai was divided between its more European western half and the more traditional Chinese eastern half. The Japanese occupation troops converted the entire city of Shanghai into a gigantic prisoner-of-war camp, closing entry and departure of its citizens.

During the Japanese occupation, the imperial army intelligence officers were skeptical of my father's Russian/stateless background: "Crawford . . . not Russian name! You are an American spy!" A Japanese officer chanted this accusation while interrogating my father at the infamous Shanghai Guard House. In the early phases of the Japanese occupation, while the Grand Theater was still operating, my father was ordered by the Japanese military intelligence to identify all foreign theater attendees. My father stated that he was merely an employee of an international company and could not provide such information without first obtaining permission. The officer then drew his samurai sword and swung it in a threatening manner around my father's head. However, my father refused to show any signs of fear and reiterated his company's policy as the sword swung closer and closer to his head. Following this frightening interrogation, Herman Crawford was released from the Shanghai Guard House. The next day a messenger came to invite him out to lunch at a geisha house as a guest of the interrogation officer. Apparently, the officer at the Guard House was impressed by my father's courage and respected him because he did not "lose face." The display of bravery was a significant component of the samurai culture.

During the remainder of the war, the Japanese occupation forces left us alone because at that time the Russian displaced persons did not fit into any category of "enemy." Japanese military sentries were stationed at all key intersections and bridges of Shanghai. Both Chinese and European pedestrians were obliged to bow before the sentries to pay their respects. However, my mother refused to bow to the Japanese military and choreographed elaborate strategies of avoiding these sentries by walking miles around specific intersections. In contrast, my grandmother Nina Rogova enjoyed such a spectacle and provided elaborate bows with many additional gestures. Then, in 1945, in the middle of the night, we heard someone on the street yelling to the prisoners of the camp: "The war is over! The war is over." We had survived the war and Japanese occupation despite major shortages of food and restrictions on movements and lack of employment of Europeans throughout Shanghai.

After the War

After the war, my father served a short stint (December 1945–April 1946) as a supervisor and interpreter for the Shanghai Air Depot before resuming his position as director of Shanghai movie theaters. I was enrolled in a French Catholic school, St. Jean d' Arc. This school offered curricula either in English or French. I began my education in the First Standard of their English curriculum without any earlier comprehension of the English language. My family only spoke Russian at home, and we had lived in relative isolation in the penthouse on the roof of Cathay Theater. As a result, I struggled during the first two years of primary school education in Shanghai trying to master the English language, while surrounded by an assortment of Russian, Chinese, and European classmates.

In 1949, with Chang Kei Sheik's nationalists in full retreat, Mao Tse Zedong's communist forces approached the city of Shanghai and established the People's Republic of China. Because one of my father's employees at the theater was married to Chairman Mao's daughter, Mao offered us asylum if we wanted to remain in China. However, because of the horrendous events in Russia precipitated by the Communist Party, we decided that it would be maladaptive to stay in Shanghai. In addition, we could not get a visa to immigrate to the United States because we were classified "stateless," and at that time the only way out of China for us was through the International Refugee Organization (IRO) founded by the United Nations. Our only option was requesting asylum in a DP camp in the Philippine Islands. Philippine President Quirino had offered refuge on a small island, Tubabao (off the South Eastern point of the Island of Samar), for 6,000 Russians escaping from China. My father arranged for our family to fly out of Shanghai on the last airplane prior to its occupation by Mao's communist forces. It was a gray, drizzly morning in mid-November, 1948, when a group of Russian emigres lined up to board the DC-6 propeller-driven airplane at the Shanghai airport. In the distance we could hear the rumbling of heavy artillery. This was my first flight in an airplane, which we laughingly termed the "vomit comet." The flight went through a horrendous typhoon, and I experienced my first airsickness, turning a nice shade of green while disposing of an earlier meal.

Tubabao, Philippine Islands

We arrived in Manila at night and discovered that no overnight indoor accommodations were available at the airport while the flight to Samar Island (next to the DP camp location) was scheduled for the next day. That night, we slept outdoors in the humid, tropical heat, without mosquito netting and thus were introduced to the mosquito population of Manila. My mother spent most of the night awake, trying to discourage mosquitoes from feasting on Marion and me. The next morning, we flew to the island of Samar and disembarked from the airplane to view an unfamiliar, iron-rich, bright-red mud and palm trees surrounding the airport. We boarded jeeps and were driven to the tiny island of Tubabao, connected by the Timbre Bridge to the southeast coast of Samar. There we were unceremoniously dumped (luggage and all) in a clearing surrounded by palm trees and tropical forest. The International Refugee Organization (IRO) of the United Nations provided US army surplus tents and tinned food. Ditches, excavated by hand, served as communal latrines with clusters of huts constructed around them. These outhouses contained five or six holes for deposition of excrement. At night, it was frightening for a young man to squat over one of these latrine holes, while a pair of shining eyes of creatures (such as lizards) were peering from the bottom of the latrine. As it turned out, this experience with communal latrines prepared me for the "culture" associated with similar facilities in Siberia. The Siberian outhouses also contained five or six holes in the wooden plank floors with no partitions and as many individuals squatting. When you enter this facility, you encounter friends, neighbors, and colleagues squatting at their respective holes while clutching the latest issue of *Pravda*, their version of the US Sears catalogue, used for wiping rear ends. Siberian culture dictates that you do not recognize any of these "squatters" and proceed to do your business without greeting anyone with a cheerful "Good morning, Ivan." In these outhouses, individual privacy was approximated through this "feigned" anonymity.

The IRO had been misinformed by the Philippine government that the former naval station on the island of Tubabao contained adequate living facilities with electricity, running water, residential huts, and roads. In reality, there was neither running water nor electricity in the camp. Initially, to get water, we had to slide down a muddy hill to a small stream, fill our buckets with dirty water, and navigate the slippery slope lugging the heavy buckets. Eventually, my father (a trained engineer) led an expedition to an abandoned US naval base and repair facility, Manicani, on eastern Samar to strip off the available pipping, pumps, and other equipment necessary for the installation of a water system. After the installation of a diesel-powered pump and an elevated sedimentary tank, we had access to showers and drinking water. Eventually, generators for the production of electricity were installed on the island of Tubabao, and each tent was wired for electricity but limited to the use of a single 25 watt light bulb for the night. Electrical power, switched off at 10:00 pm on a weekday basis and at 11:00 pm on Saturdays, allowed an extra hour for "relaxation." Nash (2002) in a book published in New South Wales, Australia, described the living conditions that one Russian family (the Tarasovs) experienced on Tubabao and discussed family immigration from the Philippines to Australia.

In the Tubabao camp, food was prepared in communal kitchens with cooking responsibilities shared by the DP women on an alternating basis. These kitchens prepared our daily rations, with "delicacies" such as hash (canned corned beef and potato mixture) and other canned meats often served in a soup or a gruel. Since the communal kitchen only served two meals a day, breakfast consisted of a slice of bread, with a cheese spread and a cup of instant coffee. Periodically, my mother was able to supplement this unappetizing diet with an occasional fresh fish purchased from Filipino fishermen and tasty mangos sold by the locals but requiring US currency. In this way, my early experiences associated with living in a tent for nine months, under tropical conditions, with an assortment of insects (mosquitos, scorpions, poisonous centipedes) and venous snakes prepared me for the future rigors of field research in a variety of "exotic places." This Philippine experience reinforced my hatred of insects and made my field research in the Aleutian Islands particularly enjoyable with no mosquitoes, a cool environment, and a highly friendly community. In contrast, working in Siberia during the summer months meant clouds of ravenous Anopheles mosquitoes and having to wear mosquito netting at all times, making outdoor meals particularly challenging (see Figure 7.4 in Chapter 7). Since electricity was periodic and refrigerators and freezers unavailable, the food stored underground in permafrost after a few days began to rot. With this unpalatable diet, I managed to lose 25 pounds in one month in the field. Later in my life, I was tempted to establish Siberian "fat farms" for those desperate to quickly slim down by refusing to eat rotting food and thus losing large quantities of fat.

Since most of the instructors in the Tubabao tent schools were from Shanghai and mostly of Russian origin, the only available schooling for children at the DP camp was in the Russian language and based on the Cyrillic script. Marion and I only spoke native Russian at home, but all instruction at our respective Catholic schools of Shanghai was in English. We chose not to attend these Russian tent schools on Tubabao because we were informed by the camp administrators that our stay in the DP camp was to be of short duration (an estimated maximum of four months), and changing the entire school curriculum from English to Russian could complicate our learning particularly in basic sciences and humanities. I could neither read nor write using Cyrillic until my undergraduate days at the University of Washington, Seattle, when I enrolled in courses on Russian literature and grammar.

Uranquinty, Australia

After nine months in this tent purgatory, with dozens of the elderly Russians succumbing to a wide variety of tropical and chronic diseases, Australia offered sanctuary. Russian Orthodox Archbishop John of Shanghai negotiated the relocation of the Russian colony from Tubabao to Australia. We boarded a US naval troop carrier, the "Marine Jumper," which sailed from Samar, Philippine Islands, to Sydney Harbor, Australia.

From Sydney, my family was transported by train across the Blue Mountains to a DP camp, Uranquinty (aboriginal term for yellow box tree and plenty of rain), 15 kilometers south of the town of Wagga Wagga, New South Wales. This large riverine town was named after the Wiradjuri aboriginal word *Wagga*, meaning "crow." Therefore, Wagga Wagga was a place of "many crows." This camp was established in 1948 primarily for East European migrants to Australia on the site of the former RAAF Air Service Flying School, 5 kilometers northwest of the township of Uranquinty. From 1948 to 1952, 28,000 displaced persons passed through this camp. Adult males at the camp were required to sign contracts for physical labor on projects at the discretion of the Commonwealth Department of Immigration. Although my father was an electrical engineer with exceptional mechanical skills and vast experience in massive construction projects and maintenance of equipment, he was contracted by the Australian government to dig ditches at a construction site in Sydney. This migrant camp in Uranquinty consisted primarily of displaced persons of European origin, transplanted by World War II, plus Russians who escaped from China through the Philippine Islands. The accommodations, consisting of wooden, military barracks, were more comfortable than the tents of Tubabao. At least we had toilets in the building instead of smelly outhouses. After one month at Uranquinty, my father began working in Sydney and was able to rent an apartment for the family on Roscoe Street, two blocks from the renowned and scenic Bondi Beach.

I was able to start schooling at Bondi Primary School, which was located approximately one mile from the apartment that we rented. Every weekday morning, I walked that mile.

Immigrants to Australia had to sign contracts to assist the development and construction of their newly adopted country by providing services in physical labor, such as ditch digging. Despite my academic interests and performance, after graduating from my primary school at Bondi, I was assigned to a technical school (Bondi Technical High School) with an emphasis on woodworking and metalworking, which I detested! Marion registered at Dover Heights High School, where she was required to take appropriate courses, such as home economics, a prerequisite for future Australian housewives. However, my academic interests lay in medicine and biological research, which required university training and attendance at a preparatory high school. At that time, some Australian educators believed that "New Australians" were not appropriate material for their universities and so to get better educational and work opportunities, my family emigrated from Australia to the United States in 1952.

Seattle, Washington

My aunt Tamara (my father's sister) and her American husband, Johnny Muttart (originally from a small community of Gold Bar, Washington), residents of Seattle, Washington, sponsored our visa from Australia to the United States. In 1952, we sailed from Sydney Harbor to Vancouver, Canada, with stopovers in Auckland, New Zealand; Suva, Fiji; Honolulu, Hawaii; and Vancouver Island before arriving in Vancouver, British Columbia. After my family disembarked from our ship at the docks of Vancouver, we cleared customs and were greeted by my aunt and her husband. They drove us by car across the border to Seattle, Washington. We lived with my aunt and her husband for two weeks before my father was able to find employment at Frederick and Nelson department store as a clerk unloading merchandise at the dock. We were then able to first rent a house located a few blocks from the Muttart residence in Ballard, Seattle, and eventually buy a house a few miles away.

In Seattle, my parents enrolled me at a Catholic high school, O'Dea, despite not being of Catholic persuasion and becoming an eventual atheist! Several of my uncle's relatives (particularly his nephew, Jack Anderson) attended Catholic schools in Seattle and attested to the high quality of education. I started high school during the month of October 1952, at the tender age of 13 years, in part because of my academic record, letters of recommendation from the Bondi Technical School, and because of the different organization of schooling in Australia. At that time, Australian children were required to obtain six years of primary school, followed by three years of regular high school; and if the students showed academic potential, an additional two years of college preparatory school were required. As a result, Marion completed three years of high school training by 16 years of age and then started working for a pharmaceutical company in Sydney.

I enrolled in high school in Seattle despite having missed the beginning of the official school year. As a result, I was not exposed to an introduction to the study of Latin but was given a textbook by Brother Cerasoli (affectionately nicknamed Mr. Peepers, from a popular TV program of the 1950s) and instructed to catch up to the class. Although eventually I found Latin to be extremely useful in biology, I struggled with it during my first year of high school. I also had to absorb mathematics without some of the underlying basics that my classmates had learned earlier. I was two years younger than most of my classmates at O'Dea. O'Dea provided solid training in chemistry, mathematics (algebra, geometry, and trigonometry), and history courses, but there was a lack of classes and training in biology and evolution. Thus, I had to catch up on some of these disciplines during my tenure at the University of Washington.

Undergraduate at the University of Washington

I completed my undergraduate degree in anthropology with a minor in biology in four years. Because of my lack of experience in several disciplines, my undergraduate grades were slightly above average, particularly in courses dealing with biology and evolution. I had taken a course from a young, enthusiastic anthropologist, Dr. Edgar "Bud" Winans, and he captured my interest in the field of anthropology. I had also taken some courses in general genetics, with a focus on corn and red bread mold (*Neurospora crassa*) from Professor Herschel Roman and human genetics from Professor Stanley Gartler. I found human genetics to be particularly fascinating, at least until I enrolled in a class on population genetics within the Anthropology Department. The instructor was Professor Virginia Avis, a primatologist specializing in primate brachiation (arm-over-arm swinging locomotion in trees by gibbons and orangutans). Because there was a shortage of faculty in Anthropology with any expertise in genetics, the unfortunate Virginia Avis was assigned by the department to teach in an area that she lacked any formal training. She received her Ph.D. at the University of Chicago from Sherwood Washburn, who at that time was training a bevy of nonhuman primate behavioral specialists.

First Research Experience

Dr. Bertram S. Kraus, a physical anthropologist with training at the University of Arizona, had an appointment at the Dental School of the University of Washington. His research specialty was the embryogenesis and development of the craniofacial region and dentition in humans (Kraus, 1969). I met with

Bert Kraus, and he was enthusiastic about the prospects of having a student from Anthropology working with him. He gave me three human embryos, preserved in formalin, stored in jars, and instructed me to dissect the craniofacial regions to learn about the morphology of embryos. Despite my wife's protestations, I placed these embryos in a closet but could not muster sufficient interest in their dissection. I returned the fetuses back to Kraus without dissecting any of them.

After searching the literature, I was intrigued by the apparent Mendelian bases for the ability to taste phenylthiocarbamide (PTC) and the distribution on human tongues of taste buds that sense bitterness. Kraus arranged for my use of a dental chair at the University of Washington Dental School, and I convinced dental patients to participate in this research on the localization of taste buds for bitterness. I was able to demonstrate that the taste buds (receptors) for bitter taste were located on the vallate papillae along the root of the tongue. This was my first independent attempt at research as an undergraduate student. The results were interesting but the sample sizes inadequate for publication. Shortly thereafter, Professor Kraus accepted a position as director of the Cleft Palate Center at the University of Pittsburgh. After I completed my Ph.D. in 1967, Kraus along with the eminent professor of genetics (C. C. Li) enticed me to take my first full-time academic position at the University of Pittsburgh.

Earlier, I had supplemented my undergraduate research experiences and my finances through a research assistantship with Dr. Dean Crystal, a cardiologist who at that time was developing methods of heart transplants. My unpleasant task as a research assistant was to collect dogs scheduled to be euthanized by Seattle veterinarians and to prepare these dogs for open-heart surgery. I would set up the surgical theater, load the heart-lung machine with blood, and assist in the open-heart surgery. At this time, I learned to draw blood by venipuncture, a skill that I was later able to employ in the field. This research experience convinced me that I did not want a medical career in surgery of humans.

After graduation from the University of Washington, I could not decide whether to apply for medical school or to focus my interests on academia and research. I delayed this decision by applying to the Naval Officer Candidate School in Rhode Island and spent four months in the Navy. After four months in the Officer Candidate School in Newport, Rhode Island, I washed out for health reasons and returned to Seattle, where I obtained a research assistantship with Dr. William Cantrell, chief of Surgery at Harborview Hospital in

Seattle. Cantrell was interested in the effects of hormones (particularly those associated with pregnancy) on wound healing and wanted me to develop a research program to explore these effects on animal models. I combined histochemistry with surgical procedures on the paravertebral muscles of guinea pigs and later with the insertion of corneal implants. I documented the invasion of epithelial cells to seal the excision and, through an implantation of small sponges, measured biochemical changes in levels of various markers.

Professional Beginnings

In the fall quarter of 1963, I was accepted into the graduate program by the Department of Anthropology, University of Washington. I was intrigued by how populations differentiated genetically from each other through migration and the processes of evolution. I enrolled in a seminar given by a visiting professor from Oxford University, Dr. Derek Francis Roberts, who had conducted a classical study of genetic drift in a small island population of Tristan da Cuhna, a volcanic island located in the South Atlantic Ocean (Roberts, 1968). During this seminar, he suggested that I review the literature on nonhuman primate genetic variation based on proteins and present it to the class. At that time, the method of starch gel electrophoresis, developed by Orville Smithies (1959), was being applied as a rough measure of genetic variation based on differential mobility of protein variants on a starch gel medium. It took two consecutive three-hour seminar sessions to review the vast new electrophoretic and immunological literature. After I completed this review, Roberts prompted me to write it up and submit a manuscript as my M.A. thesis. Unfortunately, because of a conflict between professors Derek Roberts and social anthropologist Kenneth E. (Mick) Read, chair of Anthropology, Roberts returned to the United Kingdom.

Fortunately, the teacher who introduced me to human genetics, Dr. Stanley Gartler, agreed to supervise my M.A. thesis: "A Re-evaluation of the Taxonomy and Phylogeny of the Hominoidea, Based upon Biochemical and Cytogenetic Evidence." I utilized published data from one- and two-dimensional starch gel electrophoresis, immunological methods (immunodiffusion), red cell blood groups, and cytogenetics to evaluate phylogenetic reconstructions previously based on morphology of the Hominoids.

The summer of 1964 was particularly challenging. I applied and was accepted into a Summer Institute on Behavioral Genetics at the University of

California–Berkeley sponsored by the National Institutes of Health (NIH). In addition, I also received financial support from Wenner-Gren Foundation for Anthropological Research to attend the VIIth International Congress on Anthropological and Ethnological Sciences in Moscow, USSR. This was the first international, anthropological Congress held in the Soviet Union. Wenner-Gren Foundation sought graduate students with knowledge of the Russian language, who could serve as translators at informal sessions between Russian and American scholars—away from the prying eyes and KGB microphones. What made this fellowship so attractive to a graduate student was that it provided a glimpse into the ongoing transition from Stalinist to "Kruschevian" science in Russia and its impact on European academics.

Behavioral Genetics Institute

The six-week Behavioral Genetics Institute ran from June 23 to July 31, 1964, and featured an assortment of eminent geneticists, physiologists, anthropologists, psychiatrists, and biologists. It also included visits to the Sonoma State Hospital and Palo-Alto Veterans Administration Hospital for direct observation of genetically influenced mental illnesses such as schizophrenia and manic-depressive disorders. The program at the Institute included lectures on topics associated with behavioral genetics by internationally eminent scholars. The lecturers included David Rosenthal (genetics of schizophrenia); Jerry Hirsch on behavioral genetics of fruit flies and selection for specific behaviors; D. J. Merrell on the mapping of chromosomes and gene action; James McGaugh on stimulus-response in animal models; and Cradock Roberts on quantitative genetics and measurement of heritability. I was fortunate to be able to discuss evolutionary theory using the Russian language with Professor Theodosius Dobzhansky, one of the founding fathers of the modern evolutionary synthesis. This Institute generated new perspectives for me on the world of genetics—not only on mechanisms of behavior but also population and quantitative approaches to genetics.

World Congress of Anthropological Sciences—Moscow

My next academic adventure began on July 30, 1964, with flights from Berkeley to San Francisco to Dulles (Washington, DC) to London, UK, to Warsaw,

Poland, and on August 2 to Moscow, USSR, for the VII World Congress on Anthropological Sciences. This was a heady experience for a young, aspiring biological anthropologist from the University of Washington. I met an array of well-known anthropologists, namely James Spuhler (early anthropological geneticist from University of Michigan); Henry Field (research fellow at the Field Museum of Natural History); and Mikhail Gerasimov (Russian archaeologist and anthropologist), who reconstructed the faces of Neandertals, Java Man, Tamerlane, and Ivan the Terrible (Ivan Groznii). Yuri Rychkov (population geneticist), despite the official Lamarckian policy in the USSR, applied the modern synthesis to the study of evolution. I was also able to discuss potential research collaborations with members of the Institute of Experimental Pathology and Therapy Center from Sukhumi, Abkhasia, and with Russian academics from the Anthropology Institute at Moscow's State (Mikhail Lomonosov) University. I experienced the richness of Russian/Soviet Union culture by attending the Prokopieff's opera *Love for Three Oranges*, Tchaikovsky's opera *Evgenie Onegin*, a Moscow Symphony Orchestra concert, and the Georgian National Dance Company performance at Gorky Park. In addition, I was fortunate to visit a number of exceptional museums: Pushkin's Historical and Art Gallery, Historical Museum on Red Square, Museum for the Development of Religion and Atheism in St. Petersburg, and Ethnographic Museum of Latvia (Riga). It is through the numerous professional contacts and friendships acquired during this Congress and after that I was able to establish long-term research programs in Siberia and in Sukhumi, Abkhasia. During the 1964 travel, I did not get an opportunity to meet with Boris Lapin (director of the Primate Center in Sukhumi); however, we did establish a project when he later visited the United States to inspect the newly established Regional Primate Center at the University of Washington, Seattle.

Spuhler recommended that I apply to the University of Michigan for a Rackham fellowship to complete my Ph.D. At that time these fellowships provided only $200/month stipends, and my family was about to increase with Carolynn, my wife, expecting a child, in November 1964. Therefore, we could not afford to move to Michigan and subsist on $200/month. I made an appointment with the director (Dr. Ted Ruch) of the newly established Regional Primate Research Center at the University of Washington, and he suggested that I outline a potential program for nonhuman primate genetic research. Within a few days, I submitted this document of potential research programs and was appointed (at $600/month), to serve as the sole primate geneticist of the Regional Primate Center, University of Washington. Dr. Ruch assigned to me a state-of-the-art biochemistry laboratory at the Primate Center and provided funding for laboratory supplies, equipment, and professional travel to national and international meetings. I also had access to an editor, a photographer, and a first-rate library on primate research. Because of such an attractive position, I was able to complete my research and dissertation in two years and publish, as sole author, an article in the prestigious journal *Science* (Crawford, 1966).

Ph.D. Dissertation

My doctoral dissertation research focused on the phylogeny of the Hominoidea, based on biochemical and immunological markers (Crawford, 1967). The purpose of this study was twofold: (1) to determine the amount of genetic variation in the Hominoids based on eight blood loci; and (2) to evaluate theories of hominoid phylogeny and taxonomy. Protein and immunological genetic markers were utilized to explore primate taxonomy and phylogeny. At that time, the taxonomy of the Great Apes and humans was primarily based on morphological similarity and differences. I obtained from eight regional primate centers and a number of zoos a total of 132 blood samples representing the major hominoid taxa. Eight protein systems from the erythrocytes and serum were characterized using starch gel electrophoresis. These loci included glucose-6-phosphate dehydrogenase (G-6-PD), 6-phospho-gluconate dehydrogenase (6PGD), serum pseudocholinesterase, acid phosphatase (AP), phosphoglucomutase (PGM), hemoglobin (Hb), transferrins (Tf), and haptoglobins (Hp). Several immunological techniques (micro-complement fixation, immunodiffusion, and percentage fixation) were utilized to test genetic affinities among representatives of the Hominoidea. These experimental data demonstrated that the Chimpanzees (Pan) were phylogenetically more closely related to humans than to any other primate. The Gibbons (Hylobates) belong to a separate subfamily within the family Pongidae. Many of the ambiguities associated with the application of electrophoresis and immunological techniques to taxonomy do not exist now that DNA sequencing is available and employed directly for phylogenetic reconstruction.

Because of the rapid turnover of Physical Anthropology faculty at the University of Washington, I was trained in both traditional anthropology by

Franz Boas's former students (Mel Jacobs and Vern Ray) and modern genetics by Arno Motulsky, Derek Roberts, and Stanley Gartler. In addition, I can trace my intellectual pedigree to Albert Hooton, through Marshall Newman and Virginia Avis. In 1966–1967 during my last year in graduate school, Marshall (Bud) Newman accepted the biological anthropology professorship at the University of Washington. He assumed the chair of my doctoral committee because Arno Motulsky (former chair) was on leave, conducting field research on the genetics of malaria resistance in Central Africa. Motulsky, an MD, had done a postdoctoral fellowship with James Neel at the University of Michigan. Neel introduced Motulsky to biochemical and medical genetics. Virginia Avis completed her dissertation on primate brachiation with Sherwood Washburn, a former student of Hooton. I also had a course on evolutionary theory from Theodosius Dobzhansky and an assortment of geneticists and psychologists at the Behavioral Genetics Institute at the University of California–Berkeley. Derek Roberts traced his academic roots to eminent statistician R. A. Fisher at Oxford University. I acquired first-hand biochemical laboratory skills from Professor Akira Yoshida, who supervised and suffered through my attempts to purify red blood cell enzyme glucose-6-phosphate dehydrogenase (G-6-PD) on a chromatography column. I managed to get formal training in biochemical and population genetics from a series of mentors, while remaining at the University of Washington (see Figure 2.1).

I am not recommending that every reader of this volume who is contemplating field research in another country should experience an identical background to mine. To successfully complete fieldwork in another country, you need not necessarily be born in another country, such as China, survive several DP camps in the Philippine Islands, and live in a tent under tropical conditions for one year. Instead, traveling and having broad cultural experience and interacting with people of different social and economic backgrounds will help you adapt more easily to field conditions. Learning the languages of the populations that you want to study can be of great assistance as well. In my case, being able to speak Russian was of great help in initiating the Siberian research.

Genetically, I consider myself a case of cultural and genetic "heterozygosity," that is, with a combination of several ethnicities: Scottish, Russian, Buriyat, and German ancestry. In addition, I was raised in an assortment of environments, such as urban China, tropical Philippine Islands, temperate Australia, and the United States. This complex of genetic/environmental interactions prepared me for a career in anthropological genetics and a fondness of field investigations.

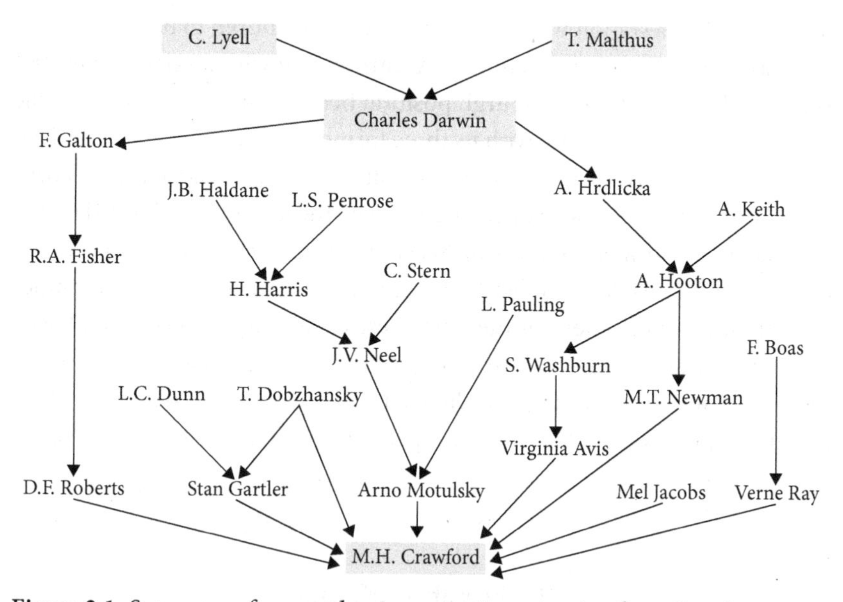

Figure 2.1 Summary of my academic connections, tracing from Derek Roberts back to R. A. Fisher and Charles Darwin. My training and research collaboration with Arno Motulsky gave me an indirect connection to James Neel and Curt Stern.

Professional Experiences: Navigating the Choppy Waters of Anthropology

University of Pittsburgh

During the summer of 1967, I defended my doctoral dissertation before the faculty and students of the Department of Anthropology, at the University of Washington. My dissertation applied protein chemistry and immunology to the taxonomy and phylogeny of the Hominoidea. Since none of the faculty within the Department of Anthropology had the background to evaluate such genetic/biochemical research, Newman sent my dissertation for review to the two most prominent specialists at that time, namely Professors Morris Goodman (Department of Microbiology, Wayne State University) and John Buettner-Janusch (Department of Anthropology at Duke University). After my dissertation was reviewed and approved by these two external reviewers, I defended it formally before the Department of Anthropology at the University of Washington in August 1967. Within a few days of the defense of the dissertation, my family (wife Carolynn and son Kenneth) and I drove in a Chevy Nova station wagon from Seattle to Pittsburgh to begin a tenure-track assistant professor position in Anthropology with a salary of $9,500/year. I had accepted the Pittsburgh position because of the reputation of the Anthropology Department with its three former presidents of the American Anthropological Association on the faculty. They were as follows: George Peter Murdoch (who developed the Human Relations Area Files [HRAF] while at Yale and was honored for his contributions to science by Hirohito, emperor of Japan); John Gillen (Latin American specialist); and Alexander Spoehr (former president of the East-West Center at University of Hawaii and a renaissance archaeologist). Despite my youth and professional inexperience, I was appointed by the University of Pittsburgh to develop a physical anthropology presence in what appeared to be a first-rate Anthropology Department.

University of Kansas

I had interviewed at several other universities and research institutions before settling on Kansas. These included the University of Illinois at Urbana, CUNY-Hunter College in New York City, the New York Blood Bank, and the University of Wisconsin–Madison. After relocating in Lawrence, I was recruited for the chairmanship of a highly divided and combative Anthropology Department (consisting of Maoists, political conservatives, and a physical anthropologist, Bill Laughlin, who was isolated like some virulent bacterium at the Medical Center) at the University of Connecticut in Storrs. However, since I had received a Research Career Development Award in 1976 from NIH, which freed me to focus on my research, I rejected the challenging offer from Connecticut.

Although my academic specialty was anthropological genetics, I was appointed by the Department of Anthropology, University of Kansas, as a replacement for the eminent forensic osteologist Professor William Bass.

Research at the LBA

In 1976, I received a five-year Research Career Development Award from NIH and since this grant freed me from teaching duties for five years while providing a 12-month salary, I decided to stay at the University of Kansas.

This Career Development Award freed the funds from my nine-month salary, normally provided by the state of Kansas. These funds permitted the university to hire my replacement for three years and then to support an international lecture series for two additional years that resulted in the publication by Plenum Press of two volumes on *Current Developments in Anthropological Genetics* (Mielke and Crawford, 1980; Crawford and Mielke, 1982). I added a newly hired junior faculty member, James Mielke, as senior editor of the first volume of the Plenum series. In addition, we collaborated on a study of the genetic structure of Inuit populations, published in the *American Journal of Physical Anthropology* (Crawford et al., 1981). A third volume in the Anthropological Genetics series published by Plenum Press was built around my research program with the Black Caribs of Central America and the Caribbean and was independent of the University of Kansas lecture series (Crawford, 1984). Most of the contributors to volume 3 (Current Developments in Anthropological Genetics, Black Caribs: A Case Study in Biocultural Adaptation) were participants in an all-day symposium that I organized and held at the American Anthropological Association meetings in Houston in 1977.

The funding of my grant applications by National Institute of Dental Research (NIDR), National Cancer Institute (NCI), National Institute of Aging (NIA), National Science Foundation (NSF), and the US Office of Education provided considerable leverage for the establishment of the Laboratory of Biological Anthropology (LBA) in 1975 (Crawford, 2016). The availability of such facilities plus generous federal funding permitted the development of a number of simultaneous research programs on (1) dental evolution of transplanted Tlaxcaltecan populations of Mexico; (2) admixture estimates of Mexican Mestizo populations; (3) genetic micro-differentiation of indigenous populations of Siberia and Alaska; (4) the genetics of biological aging of Mennonite populations of Kansas and Nebraska. This Mennonite project continued for more than 30 years and developed physiological and neuromuscular measures of biological aging; (5) environmental influences affecting learning in an African American community of Kansas City. This project was developed and initiated by research assistant Amy Ferrera, one of three graduate students who transferred from the University of Pittsburgh to continue working with me at the University of Kansas; and (6) genetic epidemiological study of a lymphoma epidemic in a baboon (*Papio hamadryas*) colony of the Institute of Experimental Pathology and Therapy of Sukhumi, USSR. Two doctoral students, Dennis O'Rourke and Robert Baum, received support from NCI under a contract to the LBA and collected morphological and genealogical data for their dissertations based on the effects of inbreeding on a free-ranging sibship of baboons of Sukhumi. In addition to the availability of funding and laboratory space, three generations of graduate students and postdoctoral fellows helped generate a search for human evolutionary effects on morphology, population genetics, and disease. Table 12.1 of this volume summarizes the chronology, locations, and sponsors of field research programs, conducted from 1968 to the present. In this book, I mainly stress my involvement in holistic programs that examine the actions of evolutionary forces and migration.

In 1989, Wayne State University Press, through associate editor Alice Nighosian, appointed me editor-in-chief of the historic journal *Human Biology*, established in 1929 by eminent biologist Raymond Pearl (Crawford, 2018). I viewed this editorial change as an opportunity to redirect the journal *Human Biology* from coverage of general human biology to a new focus on anthropological genetics. The first issue of *Human Biology* that I edited together with past editor Gabriel Lasker was a special double issue on the Foundations of Anthropological Genetics (Crawford and Lasker, 1989). We reprinted ten of the most significant publications in the field, including contributions by eminent geneticists: J. B. S. Haldane, James F. Crow, James V. Neel, Arno Motulsky, James Spuhler, Frank Livingstone, Derek Roberts, and Morris Goodman. These reprinted articles were accompanied by brief updates written either by the original authors or (if the authors were deceased) major contemporary specialists in the field of anthropological genetics. This publication provided an opportunity to trace the roots of anthropological genetics to its founders and to position the journal *Human Biology* as the leading publication in the field of anthropological genetics (Crawford and Lasker, 1989).

Professional Recognition

During my tenure at University of Kansas, my contributions to the fields of anthropological genetics and human biology (from 1971 to 2020) were recognized through the following awards by a variety of scientific organizations and institutions. This recognition made possible the receipt of more than 90 grants from national and international agencies that were used to conduct 35 field investigations.

1. American Association for the Advancement of Science (AAAS) honored me in 1996 by naming me fellow of the Association. This award was given at the AAAS national meeting in Seattle, Washington.
2. In 1995, the American Association of Anthropological Genetics (AAAG) elected me as its first president. Together with Moses Schanfield and Dennis O'Rourke, the three of us established this association at a meeting in Denver, Colorado. The primary function of AAAG was to support the journal *Human Biology* and to help develop a field that came into existence in 1973 with the publication of the volume *Methods and Theories of Anthropological Genetics.*
3. In 1997, I was elected president of the Human Biology Association. This society published a newly created journal, *American Journal of Human Biology*, a competitor of the classic *Human Biology.* My duties included president elect, 1997–1998; president, 1998–2000; past president, 2000–2001. This election placed me in somewhat of an awkward position, being the editor-in chief of *Human Biology*, while president of an organization that supports its intellectual competition.
4. The Argentine Academy of Sciences of Cordoba inducted me as a foreign academician in 2001. This induction was an honor that I shared with Charles Darwin, who accepted membership to the same academy more than 100 years earlier, on March 18, 1879 (see letter of acceptance from Darwin in Appendix A).
5. On April 14, 2011, the Human Biology Association at its national meeting honored me with the Franz Boas Lifetime Research Achievement Award. This award (presented by colleague and friend William Leonard) was given to me for lifetime achievements and contributions to the field of human biology.
6. Dennis O' Rourke presented to me the Charles R. Darwin Research Achievement Award on behalf of the American Association of Physical Anthropology in April 15, 2016.

Conclusion

Through the last 50 years of my academic career, I managed to remain relatively current in a rapidly changing scientific discipline and was able to navigate the informational and molecular revolutions through the assistance of three generations of highly talented graduate students (41 Ph.D.s mentored) and 20 postdoctoral fellows. Mentoring highly talented students not only makes the investment of time and effort feel worthwhile, but it opens up new collaborative opportunities and encouraged me to stay on the cutting edge of the field. As a result, a number of my former doctoral students have continued collaborating with me on various research programs until the present time. Appendix B lists the Ph.Ds. that I mentored from the first, Robert Halberstein (who transferred from the University of Pittsburgh with me in 1971), to the last, Randy David, from SUNY.

3

Admixture and Genetic Differentiation of Transplanted Tlaxcaltecan Populations

Introduction

Anthropological geneticists generally agree that in order to detect the actions of evolutionary forces an extended time dimension is necessary. However, historical reconstructions and major environmental changes provide insight into how evolution operates. Darwin originally viewed evolution through the lens of survival of the fittest—that is, natural selection. The modern synthesis demonstrated that four forces of evolution, namely natural selection, mutation, genetic drift, and gene flow, interact to modify the genetics of human populations. The significance of each force of evolution and their interaction depends on the effective size of the population and its demography and history. Populations that have small effective sizes and are reproductively isolated are more likely to experience stochastic fluctuations in the frequencies of genes. However, large populations undergoing migrations will exhibit allelic frequency changes due to gene flow and possibly selection. For example, Native American populations have experienced genetic changes through massive European gene flow and selection, particularly due to the introduction of new diseases. Through the study of the genetics of transplanted populations over lengthy time periods, history provides unique opportunities to evaluate evolutionary change resulting from migration and selection through measures of high mortality and low fertility rates (Crawford et al., 2021).

History of Tlaxcala, Mexico

In the 1970s, Tlaxcala was the smallest (4,600 square kilometers) and economically the poorest state in the Republic of Mexico with a population of 454,000 persons (Crawford et al., 1976). This state occupies the

In Search of Human Evolution. Michael H. Crawford, Oxford University Press. © Oxford University Press 2024.
DOI: 10.1093/9780197679432.003.0003

central and northern regions of the Tlaxcalan-Pueblan Valley in the Central Highlands of Mexico. The State of Tlaxcala is bordered on the west by the Valley of Mexico and to the east and south by the State of Puebla (see map in Figure 3.1).

While at the University of Pittsburgh, I learned from my colleague Professor Hugo Nutini (an Italian-Chilean cultural anthropologist and a former Olympic Games bronze medalist in the 800-meter run) about the unique role that the Tlaxcaltecan people played in the conquest of Mexico (Crawford, 1976). In the year 1521, the Tlaxcaltecans formed a military alliance (known as the *Segura de la Frontera*) with conquistador Hernan Cortes and the invading Spanish expeditionary force. At Contact times, the people of the Valley of Tlaxcala had been the objects of continuous hostility and warfare with the Aztecs of the adjoining valley (see map in Figure 3.1). The Aztecs did not conquer the Tlaxcaltecans but periodically raided the Valley in search of sacrificial victims for their bloody ceremonies to abate various gods.

For their services to the Spanish Crown, the Tlaxcaltecans were rewarded with a number of privileges. These included being addressed by the honorific title of "Don"; they were also permitted by the Spaniards to ride on horseback, and their land in the Valley of Tlaxcala was not confiscated by the conquering Spaniards to set up haciendas (Nutini,

Figure 3.1 Map of Mexico and the State of Tlaxcala, which locates the sites of transplantation of Tlaxcaltecans to Saltillo and Cuanalan.

Source: Reproduced from Crawford, M. H. (Ed.), *The Tlaxcaltecans: Prehistory, Demography, Morphology and Genetics*. © 1976, The Author.

1976). The Spaniards residing in the State of Tlaxcala were limited to three administrative centers. In return, the Tlaxcaltecans served as mercenary troops involved in the subjugation of Mexico. Because of this alliance with the Spaniards and the role they played in the conquest of Mexico, the Tlaxcaltecans were considered by the majority of Mexicans as traitors who sold out to foreign invaders. In 1521, a garrison from the Valley of Tlaxcala was transplanted to the adjoining Valley of Mexico to construct and protect irrigation dikes and to prevent the Aztecs from flooding the region. In 1571, the City of Saltillo was founded by conquistador Alberto del Canto. The Spanish, Native American settlement of Saltillo had been constantly besieged by the warlike Chichimec tribes of Northern Mexico. In response to these hostilities, in 1591, 400 families from the Valley of Tlaxcala were relocated to northern Mexico to establish the village of San Esteban de Nueva Tlaxcala, now known as Saltillo (Aguirre, 1976). The Spanish did this to encourage the cultivation of the land and to accelerate the stalled colonization of the region. Initially, Saltillo (now a city of more than 800,000) grew slowly because of water shortages and constant warfare with the surrounding Chichimec tribes. This fission of the Tlaxcala gene pool occurred more than 350 years ago with the settlement of the families from the moist, temperate altiplano (high mountainous plateau) of Central Mexico to the arid, desert-like landscape of northern Mexico.

John McCullough (at that time a graduate student from Penn State University who was surveying the Valley of Mexico for possible archaeological sites) informed me that in 1521 a garrison from the Valley of Tlaxcala was transplanted to the adjoining Valley of Mexico. This enclave of Tlaxcaltecans became a barrio of Cuanalan, located within the municipio of Acolman (see Figure 3.2). While the exact size of the original garrison was not historically known, the earliest census from the 18th century enumerates a total of 212 (Indian) "Indio" and 10 Mestizo families. This barrio is within sight of the ceremonial complex of Teotihuacan, with its pyramids and temples (see Figure 3.2). Social isolation, due to the perceived view of most Mexicans that the Tlaxcaltecans were "traitors," made a genetic analysis of the contemporary descendants of the garrison of potential evolutionary interest. The experimental design for this study consisted of an assessment of genetic, morphological, and demographic differentiation between the populations within the Valley of Tlaxcala (San Pablo del Monte and the City of Tlaxcala) with two transplanted populations, Cuanalan and Saltillo. Given more than 350 years of separation, have these groups

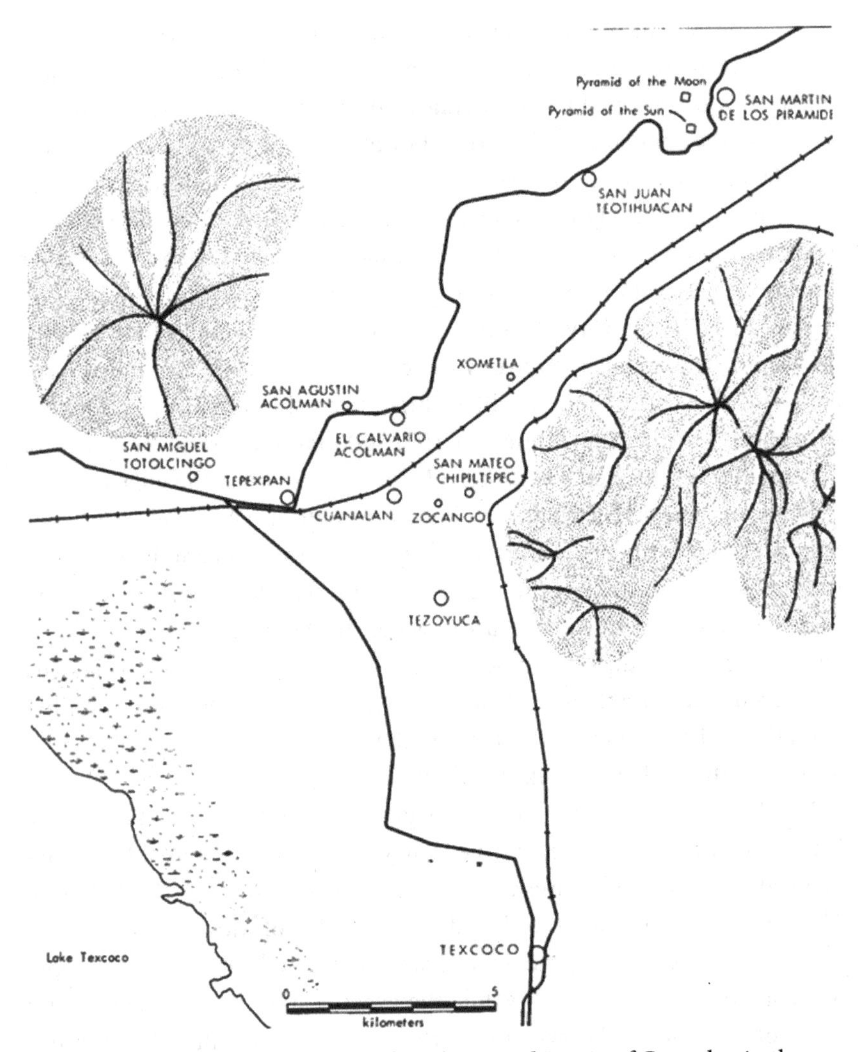

Figure 3.2 Map of the Municipio of Acolman and Barrio of Cuanalan in the Valley of Mexico.

Source: Reproduced from Crawford, M. H. (Ed.), *The Tlaxcaltecans: Prehistory, Demography, Morphology and Genetics*. © 1976, The Author.

differentiated genetically, morphologically, or culturally? There is evidence to suggest that Native Americans from Tlaxcala built the earliest Catholic Church in the Americas in what is now Santa Fe, New Mexico. However, no records were located as to the exact numbers of Tlaxcaltecans brought to Santa Fe, nor any data concerning the current geographic distribution of their descendants.

Valley of Tlaxcala—Research Team

In 1969, I assembled a multidisciplinary research team primarily from the University of Pittsburgh, and we began a research program on two populations from the Valley of Tlaxcala—an indigenous Native American community of San Pablo del Monte and a Mestizo City of Tlaxcala. San Pablo was a geographically isolated community of approximately 10,000 inhabitants located on the slopes of the volcano La Malinche (named after Cortes's translator and concubine). According to several cultural anthropologists, this community experienced little, if any Spanish admixture. Genetic characterization of this village and the absence of any European marker genes supported the selection of this community as representative of Tlaxcaltecan Native Americans. The City of Tlaxcala, with a population of approximately 15,000 Mestizos, was founded in the 16th century as an administrative center by the Spanish and as a result reflected considerable Spanish gene flow into the contemporary population.

Early Human Experimentation

At the time of the development of the Tlaxcala research, the US governmental regulations on human experimentation were undergoing major modifications through the creation of Internal Review Boards (IRBs) and the monitoring of risk in research. This research program in Tlaxcala preceded the enactment of the National Research Act (Pub. L 93-348) by the US Congress in 1974, which mandated that IRBs review all Public Health Service–funded research. This committee was charged with the identification of the basic ethical principles that should underlie the conduct of biomedical and behavioral research involving human subjects. However, the Tlaxcala field research program was initiated before the documentation of ethics violations associated with the Tuskegee study of syphilis in African Americans and the implementation of laws and regulations associated with human research.

In the early 1970s, the Belmont Report (1974) on ethical guidelines for fieldwork had not yet been compiled or published. Although the IRBs had not been constituted, by the time this research program was organized, we followed the general guidelines for human experimentation as outlined by distinguished geneticist and physician James Neel and Salzano in 1964 to the World Health Organization. They stressed that local populations should be informed about the research risks, methods, and benefits from medical and dental-related services. In the Tlaxcala study, Hugo Nutini explained to the community through a series of lectures the nature and purpose of the research and that each individual had the option of whether to participate and to provide voluntary consent. In additional lectures by the Spanish-speaking members of our research team, the benefits and risks of participating in this study were discussed with the community and its leaders. In Tlaxcala, each individual gave verbal informed consent and chose to participate and to receive medical and/or dental care. A few of the participants accepted the medical care but refused to participate in the research program. This contrasts with the 1968 study of alpine villages of northern Italy, where the communities agreed to participate but only in the demographic/genealogical portions of the study but refused to give blood for a genetic component, thus restricting us to collecting demographic data from church records and family interviews but cancelling any genetic data collection (see Chapter 9).

Participants in Tlaxcala

The multidisciplinary research team first conducted field investigations in the State of Tlaxcala (at San Pablo del Monte and the City of Tlaxcala) for two months during the summer of 1969. The team consisted of the following researchers:

1. Dr. Seishi W. Oka, a clinically trained dentist and member of the Human Genetics Program at the University of Pittsburgh, provided dental care to participants of the study and documented variation in dentition by taking dental impressions and casts. While in the field, he also provided transportation in his Volkswagen microbus, sometimes frightening the passengers and local drivers by exhibiting a bottle of tequila as he sped through the traffic in the "Glorietas" of Mexico City.
2. Dr. T. Aidan Cockburn, originally from the United Kingdom, was the chief medical officer for the city of Detroit. He was one of the pioneers in the field of paleopathology (Cockburn, 1963). He provided medical care and obtained blood samples through venipuncture.

3. Robert A. Halberstein (better known as Halfstone or Chick Hall), a graduate student in physical anthropology at Pittsburgh, collected demographic and genealogical information by administering demographic proformae and interviewing participants (Halberstein and Crawford, 1973). He preceded the arrival of the research team by participating in a summer field school in the Valley of Tlaxcala, organized by Professor Hugo Nutini. In 1973, Halberstein went on to a Ph.D. at the University of Kansas, and eventual professorship and chair of the Department of Anthropology at University of Miami in Coral Gables.
4. Patricia Potrezebowski, graduate student in human genetics at the University of Pittsburgh, completed her Ph.D. with Professor C. C. Li and eventually became executive director of the National Association for Public Health Statistics and Information. Her research duties in this study included the preparation of blood specimens for shipment to NIH and the collection of finger and palm prints from 307 individuals of San Pablo and Tlaxcala.
5. Paul Scuilli, graduate student in physical anthropology at the University of Pittsburgh, assisted Dr. Oka in casting the dental impressions. He completed his Ph.D. at Pitt and went on to a highly successful career as professor of anthropology and chair at Ohio State University.
6. Professor Hugo Nutini, a highly experienced cultural anthropologist, who had conducted numerous field investigations in the Valley of Tlaxcala, obtained local permissions and oral informed consent from participants of this study
7. Michael H. Crawford obtained grant support for this project, organized the research stations at the public health clinics for data collection, and conducted anthropometric measurements on consenting research subjects.

Most of the team accompanied Dr. Seishi Oka, who drove his Volkswagen microbus, loaded with equipment and paperwork, all the way from Pittsburgh to Tlaxcala and back. This Volkswagen microbus was our primary means of transport in Mexico City and in Tlaxcala. Dr. Aidan Cockburn flew solo from Detroit to Mexico City while attempting to import the medications, donated for the study. Since he spoke no Spanish, he could not explain the reasons for bringing all of these "drugs" to Mexico, and the customs officials confiscated a bag full of medications. After escaping customs, he hailed the first taxi he could find at the airport and drove all the way to the adjoining valley. Aidan explained to a taxi driver that he wanted to be driven to . . . "Tlaxcala, Tlaxcala." After he greeted me in the hotel in the city of Tlaxcala, I learned about his adventures and compensated him for the horrendous cost of that taxi trip from one valley to the adjoining valley.

Methodology—First Phase

A total of 395 research subjects (138 from the City of Tlaxcala and 257 from San Pablo del Monte) participated in the initial phase of this study (Crawford et al., 1979). The population from the Native community, San Pablo del Monte, enthusiastically agreed to participate in this research. Dr. Aidan Cockburn, team physician, provided medical treatment, whenever possible. Dr. Seishi Oka provided dental care and made casts of teeth using amalgam and dental stone. Participants had an option to avail themselves of dental and medical services or could reject any participation in the research project. Some of the participants insisted that blood be drawn from both arms to drain the lungs of bad blood, and they even offered to pay the nurse phlebotomist for drawing a second sample from the other arm. I first noticed scars on both arms and traced this enthusiastic support of the bilateral venipuncture to a local curandero who punctured the arm with a sharp knife, drew a sardine can filled with blood, prayed, and then buried the can as a treatment for specific ills. I apprenticed myself to this curandero, who displayed exceptional anatomical knowledge and managed to combine bloodletting (practiced by Central American Natives by painfully inserting cactus spines into their tongues) and Spanish bloodletting as a treatment method. The curandero presented the bloodletting practice within a religious, Catholic context and the recitation of appropriate prayers accompanying the burial of a sardine can full of blood.

Initially, the Mestizos from the City of Tlaxcala were suspicious of the "gringo doctors" and were reluctant to participate in any research projects. Despite highly informative talks given to the community by Professor Hugo Nutini, explaining our research and the availability of free medical and dental services, few families initially came to our clinic. To stimulate the interest of the community, I turned out for the City of Tlaxcala football (soccer) team—surprisingly named after the Italian town of "Padua." The yellow-uniformed Padua soccer team played in a spacious stadium with well-manicured grass on the field. Following the footsteps of my distant relative (mother's side)

Vladimir Nabokov, eminent writer, and university professor at Cornell University, I played both soccer and tennis in college. However, unlike him, I played as an attacking midfielder or in a central forward position. While in exile in Germany, Nabokov played goalkeeper on the soccer team and gave tennis lessons (Field, 1977). After college, I continued competing in semi-professional leagues of the Northwest and in the Keystone league of Pittsburgh. My competitive instincts, initially displayed on the soccer field and the tennis courts, were further expressed in later academic wars. Despite the skills that I acquired while playing soccer for many years in the United States, Australia, and China, practicing and playing soccer at an elevation of 8,500 feet was a challenge in regard to hypoxic stress. I made the team and received some recognition from the community after scoring a hat trick (three goals) in a single game against a village team. Because I was stockier than most of the slight Tlaxcaltecan players, I received the nickname of "El Gordo" (fat boy). Because of my newly acquired reputation on the soccer field, families of the players started attending our clinics as the word spread throughout the community. One could say that in Tlaxcala: "I sacrificed my shins for science."

Results

The blood samples drawn by Dr. Cockburn and a nurse phlebotomist from the public health clinic were centrifuged and prepared for shipment by Patricia Potrzebowski. These blood samples, packed in ice, were sent for analysis to the Human Genetics Branch of NIDR. Blood groups, serum, and red blood cell proteins for the two populations and were analyzed by NIH laboratory technician Chad Leyshon and compared to other Native American, African, and Spanish samples (Crawford et al., 1976). We estimated triracial ancestry using multiple regression analyses on 26 allelic frequencies of the Mestizos from the City of Tlaxcala (Crawford et al., 1976). The analyses of genetic ancestry, using blood group and protein genes revealed that 16% of the genes of Tlaxcaltecan Mestizos were of Spanish origin, 76% were Native American, and 8% were of African ancestry (Crawford et al., 1981). This research preceded the molecular revolution; thus, we were limited to Mendelian markers such as blood groups and protein variants identified through blood typing and starch gel electrophoresis. Later, these initial estimates of admixture for Mestizos from Tlaxcala were compared (based on HLA molecular markers) to the ancestry of 181 participants from the City of Tlaxcala by a Mexican research team (Pavon-Vargas et al., 2019). Although the DNA markers are much more informative, a recent analysis based on HLA class I and II frequencies yielded almost identical estimates of ancestry computed by maximum likelihood (ML): Native American 75.13% +/– 1.56; European 16.10% +/– 4.98; African 8.78% +/– 4.09. The 8% African admixture among the Tlaxcaltecan Mestizos reflected the presence of approximately 3,000 Africans registered in Tlaxcala by 1580 (Aguirre Beltran, 1944). The highest incidence of African genetic markers came from populations residing along the east coast of Mexico where slaves worked on the fruit plantations and mines. If a bi-population model (based only on Spanish and Tlaxcaltecan ancestral allelic frequencies) was employed to estimate Mestizo ancestry, 70% of the genes are of Native American origin and 30% came from Spain. A tri-population ancestral model reveals that the African component only lowers the estimated Spanish ancestry. Was there some gene flow from the west coast of Mexico, or were we detecting a Moorish component in the conquistador army? It is clear from this portion of the Tlaxcaltecan research program that one force of evolution, namely gene flow, was the primary determinant of the genetic structure of populations from Central Mexico. The effects of selection operating on specific genetic markers were estimated by comparing the proportion of admixture in various genetic markers within the same population. However, small sample sizes and the questionable accuracy of the estimates of frequencies of the alleles in the ancestral populations complicated such analyses.

HLA—DNA

A more recent Mexican study (initiated by Dr. Rodrigo Barquera and his colleagues) of two Tlaxcaltecan populations, City of Tlaxcala, and rural village samples based on the molecular genetics of Human Leuckocyte Antigen (HLA) system provide similar estimates of European ancestry compared to our earlier studies using blood group markers (Pavon-Vargas et al., 2019; Adalid-Sainz et al., 2019). HLA class I (*HLA-A*, *-B*) and class II (*HLA-DRB1*, *-DQB1*) alleles were identified by PCR-SSP in 1,011 Mexicans from the State of Tlaxcala and those residing in the City of Tlaxcala (N = 181) and surrounding rural communities (N = 830). The City of Tlaxcala exhibits 71% Native American haplotypes and 20% European haplotypes and 4% African haplotypes. While this more recent follow-up study measured ancestry in rural Tlaxcala but did not sample San Pablo del Monte, which had not exhibited any European markers in the earlier study—that is, displayed 100% Native American markers. A more sensitive series of markers, HLA, detected 16% European and 4% African ancestry in the City of Tlaxcala. In the 1969 study, the village of San Pablo del Monte was much more isolated genetically and geographically from other Tlaxcaltecan populations. We had originally hypothesized that the small African component in Tlaxcala came with the Spanish army and reflected North African (Moorish) influence. However, after the HLA analysis, it is more likely that the observed African ancestry was gene flow introduced from the east coast of Mexico, where slaves were located to work the plantations.

Cuanalan—Transplanted Population

During the 16th century, one garrison from the Valley of Tlaxcala was relocated to Cuanalan (a barrio in the municipio of Acolman, located in the Valley of Mexico) to construct and guard an irrigation dike on the edge of Lake Texcoco (see Figure 3.2). In 1521, approximately 16,000 warriors from Tlaxcala accompanied Cortes in the conquest of the Aztecs in the Valley of Mexico. This region of the Valley of Mexico had been continuously inhabited for an extended period of time. There is evidence of human habitation in this region dating back to 12,000 to 14,000 years in the person of a Tepexepan pre-Columbian skeleton discovered in 1947 on the shores of Lake Texcoco.

I learned from John McCullough, a graduate student at Penn State University, who was surveying the Valley of Mexico for possible excavation sites, that in 1521 a garrison from the Valley of Tlaxcala was stationed in Cuanalan. In the summer of 1972, a field team from the University of Kansas was assembled to investigate the genetic micro-differentiation of descendants of the garrison sent from the Valley of Tlaxcala to the adjoining Valley of Mexico almost 400 years ago.

Participants

This research team from the University of Kansas consisted of the following:

1. Robert Halberstein, a graduate student in biological anthropology, participated in the 1969 study in the Valley of Tlaxcala. He collected

demographic and genealogical information in Cuanalan and wrote a dissertation in 1974 comparing the demographic characteristics of San Pablo, Tlaxcala, with the transplanted community of Cuanalan.

2. Francis Lees, a biological anthropology graduate student, wrote his dissertation on the anthropometrics of transplanted Tlaxcaltecan populations. He measured the bodies of Cuanalan research subjects and compared them to participants from the Valley of Tlaxcala and Saltillo. He later served as a faculty member at SUNY-Albany and an administrator at Rockefeller University and the Museum of Natural History in New York. In this project, Frank Lees was accompanied by his first wife, Patricia, who recorded the anthropometric measurements.
3. John Hall, cultural anthropology graduate student from the University of Kansas, is fluent in Spanish. His role in this project was to obtain verbal informed consent from the participants, to reconstruct family histories, and to explain the nature of the research to the community.
4. David Busija, a graduate student from the University of Kansas, collected dermal prints from 348 Cuanalan participants. His anthropology M.A. thesis focused on the genetics of Tlaxcaltecan dermatoglyphics. However, he completed his Ph.D. in physiology at the University of Kansas and eventually became a distinguished professor of physiology at Wake Forest University.
5. Ivanho Gamboa, a physician from Puebla, Mexico, provided health care to the people of Cuanalan. He also drew blood by venipuncture and collected information on nutrition and disease, comparing San Pablo del Monte with Cuanalan (Gamboa, 1976).
6. Ruben Lisker, a medical geneticist and physician from the National Institute of Nutrition of Mexico, obtained governmental permissions for this research program from the Ministry of Health of Mexico. Dr. Lisker and his research assistant, Rocio Perez Briceno, analyzed in the Department of Genetics Laboratory at the National Institute of Nutrition in Mexico City 536 blood samples from a total of 2,040 residents of Cuanalan (Crawford, Lisker and Perez Briceno, 1976).

Saltillo

In 1974, following the informative research on the Tlaxcaltecans of Cuanalan (from the Valley of Mexico), a research team was organized at the University of Kansas to assess the genetic and demographic changes experienced by the 400 families transplanted from the Valley of Tlaxcala to arid, northern Mexico (see Figure 3.1). Governmental permissions for this study were obtained through the Mexican collaborating physician and medical geneticist, Dr. Ruben Lisker, from the Salud Publica el Estado de Coahuila. Ruben and I first met in 1965 when he was a postdoctoral fellow working in Dr. Arno Motulsky's laboratory at the University of Washington, Seattle. At that time, I was a graduate student collaborating with Motulsky on red blood cell enzyme glucose-6-phosphate dehydrogenase (G-6-PD) variation in nonhuman primates. Field investigations in Saltillo (from July 1 to August 15, 1974) were initiated by a multidisciplinary research team from the University of Kansas (KU). We traveled in a KU state vehicle from Lawrence, Kansas, to Saltillo, Coahuila. On our arrival in Saltillo one evening, the KU vehicle was broken into and robbed of the skin color reflectometer, needles, and vacutainers. We had parked the vehicle across the street from the restaurant where we had dinner. Mistake—we thought that this was a safe parking spot because of the presence of Saltillo police. To complete the projected research, we had to purchase additional vacutainers in Mexico City.

History of Saltillo

Because of the warlike nature of the indigenous Chichimec tribes of northern Mexico, the Viceroy of Mexico recommended to the Spanish Crown that a group of peaceful, Christian, Native American farmers should be resettled in northern Mexico. Spanish administrators hoped that the Chichimecs would follow the examples of these transplanted farmers and settle to a more peaceful farming subsistence. In June 1591, an expedition consisting of 400 families (100 from each of the four administrative subdivisions of the state of Tlaxcala) and an escort of Spanish soldiers led by conquistador Zarate departed from Tlaxcala (Aguirre, 1976). The environmental contrasts between the moist, altiplano Valley of Tlaxcala at 7,500–8,500 feet elevation versus an arid, desert-like environment at 5,000 feet elevation of Saltillo should provide insight into unique evolutionary changes in the population after almost 400 years of separation. Six months later, the Spanish resettled 91 families and 16 single males (all from the principality of Tizatlan in Tlaxcala) to a new colony adjoining the beleaguered Spanish garrison of Saltillo (Nava, 1969). The remaining transplanted families from Tlaxcala were relocated to the following communities: San Luis Potosi, Colotlan, San Jeronimo del Agua, and El Venado. The City of Saltillo, being primarily of Tlaxcaltecan origin, remained of relatively small size from the 17th to 19th centuries. At the turn of the 20th century, Saltillo with industrialization experienced considerable migration and admixture from surrounding regions. By 1974, Saltillo had grown into a settlement consisting of 175,000 residents (see Figure 3.1). Given the problems of sampling such a large urban center, we decided to focus on two barrios, Chamizal and La Minita. Chamizal was an established barrio of approximately 2,000 persons, containing mostly Tlaxcaltecan surnames. In contrast, La Minita was a small shantytown with approximately 250 residents, adjacent to Chamizal and proximal to the edge of town. While Chamizal contained the descendants of the original Tlaxcaltecan settlers, La Minita represented migrants from surrounding regions who were brought from the east coast of Mexico to work in the mines of Saltillo.

Participants—Research Team

The field team of the 1974 research project consisted of the following individuals:

1. Eileen Mulhare, a bilingual Spanish-English cultural anthropologist, with her family originally from Argentina. She collected genealogical and demographic information on the community. Eileen developed into an eminent scholar of Nahua-speaking populations of the Puebla region of Mexico. Sadly, she died unexpectedly in 2006 from a heart condition followed by a massive stroke.
2. Renan del Barco, who had a career as a journalist in Lima, Peru, returned to academia for a degree at the University of Kansas in cultural anthropology. During this fieldwork in Saltillo, Renan was accompanied by his wife Delores (Dolly) and two children. He obtained informed consent and demographic and cultural data on the population of Saltillo. Renan del Barco died in 2012 at 80 years of age.
3. Delores (Dolly) Villasenor, wife of Renan del Barco, is of Mexican origin, thus a bilingual Spanish/English speaker, who assisted del Barco in collecting the genealogical and demographic data through interviews. She was accompanied by their son and daughter, Mandalit del Barco (who is currently an award-winning NPR reporter in Los Angeles).
4. Dennis O'Rourke, a graduate student at the University of Kansas. He completed his Ph.D. in 1980 and went on to a highly successful career in anthropological genetics of ancient populations. He is currently a

distinguished professor at the University of Kansas. His role in the 1974 research program was the collection of dental impressions and casts (O'Rourke and Crawford, 1976).

5. Francis Lees, a graduate student in biological anthropology at the University of Kansas. He conducted anthropometric measurements on the barrios of Saltillo and analyzed these data for his dissertation (Lees, 1975).
6. Two local physicians, Drs. Horacio Jimenez and Antonio Cardenas Alvarado, provided medical care for Saltillo residents who attended the school clinics.
7. M. H. Crawford. I obtained grant support for this project and organized the field investigations. Financial support for this study came from the National Institute of Health (NIH), the University of Kansas General Research Funds, and a Biomedical Science Support Grant.

While the exact locations of the Tlaxcaltecan settlements in Saltillo are described by historical sources, during our field investigations it became evident that the expanding business district had pushed out some of the original settlers to adjoining areas (Nava, 1969). A clinic was set up at a schoolhouse within the barrio of La Minita, two blocks from the historically known barrio of Chamizal. There was some difficulty in the self-identification of the descendants of the founders because of a stigma of being an "Indio" in this industrial, culturally Mestizo community. As a result, some of the informants were reluctant to acknowledge their Tlaxcaltecan origins. This reluctance, in its extreme form, was demonstrated by an incident in which an elderly, mustache-adorned Native American sporting a large sombrero, and being a veteran of the Pancho Villa campaign, was asked if he was a descendant of the Tlaxcaltecan founders of Saltillo. He gravely replied: "No, señor. I am not Tlaxcaltecan—but my brother is!" (Crawford, 1976).

Genetic Results

Standard Markers

Complications to the direct measurement of genetic micro-differentiation of transplanted populations, Cuanalan, and Saltillo from the Valley of Tlaxcala include the varying levels of differential gene flow and admixture. Given the perception by the surrounding Mexican populations that the Tlaxcaltecans were traitors and were socially isolated from other Mexican groups, we expected less gene flow into the transplanted groups. Earlier generations of Cuanalan residents did not experience as much gene flow, but the most recent generations intermixed with migrants heading to Mexico City.

As shown in Table 3.1, 302 persons participated in the Saltillo portion of the study, with 133 from the barrio LaMinta, 121 from Chamizal, and 48 walk-ins from adjacent barrios. This table also enumerates sample sizes for Cuanalan, San Pablo del Monte, and the City of Tlaxcala. Blood samples were drawn by venipuncture into vacutainers with ACD preservative, packed in ice and shipped for analysis directly to the collaborative Minneapolis War Memorial Blood Bank. Blood phenotyping included 12 blood group loci, 5 serum proteins, and 6 erythrocyte proteins.

Based on the allelic frequencies of 12 blood group and protein loci, a three-dimensional model was constructed based on the square root of genetic distance of Cavalli-Sforza and Edwards D^2. A ball-and-stick model reveals that the parental Indian community (San Pablo) is distinct genetically from the City of Tlaxcala (located in the same valley) and followed next by the transplanted Cuanalan residents (CUAN RES) with the least amount

Table 3.1 Sample Sizes of Tlaxcaltecan Populations Divided by Location and Migration History

Population	Sample Sizes
Cuanalan	144
Residents	205
Admixed	129
Immigrants	477
Total	
Saltillo	133
LaMinita	121
Chamizal	48
Saltillo Misc.	302
Total	
San Pablo del Monte	257
City of Tlaxcala	138
TOTAL	1174

Source: Reproduced with permission from Crawford, M. H., and Devor, E. J., "Population Structure and Admixture in Tlaxcaltecan Populations," *Am. J. Phys. Anthropol.*, 52, pp. 485–490 © 1980, Wiley.

of admixture (Figure 3.3). Of all the Tlaxcaltecan populations, Saltillo is most distant from San Pablo both geographically and genetically. This genetic distance reflects the genetic divergence due to 400 years of reproductive isolation plus some gene flow. Saltillo exhibits the closest genetic affinity to the African composite than any other Tlaxcaltecan group. San Pablo and Cuanalan residents experienced the least amount of Spanish and African gene flow.

Admixture of Transplanted Populations

Dr. Moses Schanfield (1976) analyzed the immunoglobulin haplotypes (GM and KM) for Chamizal, LaMinita, and other barrios. Chamizal reflects admixture of the Tlaxcaltecan families with the original Spanish garrison and exhibits the highest Spanish admixture of 45%, while 53% had Native American ancestry. LaMinita displays the highest African admixture

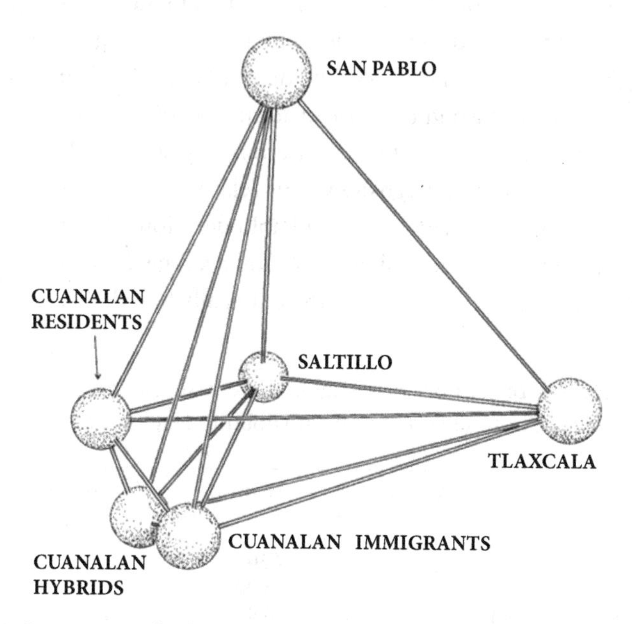

Figure 3.3 Projection of a three-dimensional model for genetic relationships between Tlaxcaltecan populations based on allelic frequencies and the square root of genetic distance D^2.

Source: Reproduced from Crawford, M. H. (Ed.), *The Tlaxcaltecans: Prehistory, Demography, Morphology and Genetics*. © 1976, The Author.

(6% based on blood group and protein markers versus 6% based on GMs) from slaves who were brought to work the mines in northern Mexico. Additionally, the degree of immigrant ancestry was confirmed through searches of the municipal records, censuses, and the administration of demographic questionnaires (Turner, 1976). Based upon family interviews, the participants from the barrio of Cuanalan were subdivided into three groups: (1) Residents are individuals with no known ancestors born outside Cuanalan for at least two generations; (2) Cuanalan Admixed (Hybrids) are persons born in Cuanalan with at least one grandparent born outside the barrio; and (3) Cuanalan Immigrants are persons born outside the barrio, whose parents immigrated into Cuanalan.

The residents of Cuanalan are predominantly Native American—96%, with 4% Spanish and with no measurable African ancestry (see Table 3.2). Those individuals who most recently migrated to Cuanalan displayed the highest Spanish ancestry: 29% (Crawford et al., 1976). Cuanalan is located a short distance from Mexico City and is a migrant stopover barrio before settlement in the big city. In Saltillo, the Residents (Chamizal) have high Spanish ancestry of 45% plus 3% African gene flow. The transplanted Tlaxcaltecans intermixed genetically with the Spanish garrison that had preceded them in the 16th century. The migrants into LaMinita worked the mines and exhibited the highest African admixture of 6%.

To assess the relative roles of stochastic versus systematic forces of evolution, mean per locus heterozygosity versus relative distance from the center (r_{ii}) of distribution were regressed. San Pablo del Monte has received much less systematic pressure (gene flow) than the average Tlaxcaltecan population. This analysis further supported the culturally based observation of the

Table 3.2 Admixture Estimates for Migrants, Admixed, and Residents of Cuanalan and Saltillo Based upon Immunoglobulin Haplotypes

Population	Native American	Spanish	African
Cuanalan			
Residents	95.70	4.30	0.00
Admixed	83.40	13.90	2.70
Immigrants	66.50	28.90	4.60
Saltillo			
Chamizal	52.50	44.70	2.80
LaMinita	57.80	36.10	6.10

Source: Reproduced from Crawford, M. H. (Ed.), *The Tlaxcaltecans: Prehistory, Demography, Morphology and Genetics*, © 1976, The Author.

absence of Spanish gene flow into the small, isolated mountainside community of San Pablo. In contrast, the barrio of Chamizal reflects greater systematic pressure, as shown by the deviation to the right of the theoretical regression line.

Serendipity and Genetics Mapping of the Rhesus Blood Group Locus

The Rhesus blood group system has been characterized by a series of five antigens (D, C, c, E, and e) in two closely linked structural genes—D and CE (Renwick, 1971). These structural genes map on the short arm p34–36 of chromosome 1. In our earlier research, we assumed (based on RA Fisher's conclusion) that the Rhesus blood group system consisted of three closely linked loci: C, D, and E. However, DNA analyses of the Rhesus system revealed that there were only two closely linked loci (Flegel, 2011). Based on blood grouping methods, a Tlaxcaltecan family was identified by a rare D- phenotype, with the presence of the D antigen but with the absence of C and E. Using cytogenetic banding techniques, a small, but significant deletion was identified on the short arm of chromosome 1 (Turner et al., 1975, 1976). These data confirmed the earlier assignment of the Rhesus complex to chromosome 1 (Marsh et al., 1974).

Statistical evaluation of cytogenetic banding and segregation of Rhesus genotypes on chromosome 1 permitted the localization of the Rhesus locus to chromosome 1. A chromosomal deletion of the A region by band pIa3 was found in three familial pedigree members of family G (Turner et al., 1975).

Morphological Differentiation of Transplanted Populations

Morphological traits, based on anthropometrics, odontometrics, and dermatoglyphics, are the result of polygenic inheritance interacting with environmental factors. Transplanting Tlaxcaltecan populations to different regions of Mexico with contrasting environments provided a unique opportunity to examine the sensitivity of the genes controlling these complex traits to environmental influences. However, there were major complications to the interpretation of the observed differentiation of the Mexican transplanted populations: (1) the degree of gene flow and admixture with Spanish,

African, and non-Tlaxcaltecan Native American; and (2) interobserver error in the anthropometric measurements.

Anthropometrics

Nine standardized anthropometric variables assessed morphological variation in 606 participants from the six Mexican groupings: Cuanalan (Residents, Admixed Immigrants), San Pablo del Monte, City of Tlaxcala, and Saltillo. The nine standardized anthropometric traits included head length, head breadth, head height, bizygomatic breadth, total face height, jaw height, nose length, and head circumference. These specific measurements were used in this study because of their higher heritability, previously observed in linear traits and facial dimensions in Mennonite populations (Devor et al., 1986a, 1986b). The age of each participant and their stature were not significantly different among males or females; however, every other variable used in the one-way analyses of variance was significant (Lees et al., 1976). Mahalanobis's D^2 genetic distance analysis for both males and females revealed that San Pablo (the Native American community) was unique when compared to the other groups. The City of Tlaxcala, located in the same valley, is closer to San Pablo del Monte since they shared similar environments. Interpopulation variation was assessed in the socially defined subdivisions of Cuanalan into Residents, Admixed, and Migrants. As expected, Cuanalan Residents clustered with both San Pablo males and females. This observation is in agreement with models predicting greatest morphological similarity between populations that share the most recent common ancestry (Lees and Crawford, 1976). Figure 3.4 is a three-dimensional scattergram of male group means revealing the optimal separation of groups along each axis. San Pablo del Monte, Saltillo, and Tlaxcala are separated along the first axis, while Cuanalan differs statistically along the second axis (Cavalli-Sforza and Edwards, 1967).

Dermatoglyphics

Finger and palm prints have been widely used for forensic identification by law enforcement agencies and anthropologists for the assessment of genetic affinities among human populations (Holt, 1968). Methods for the

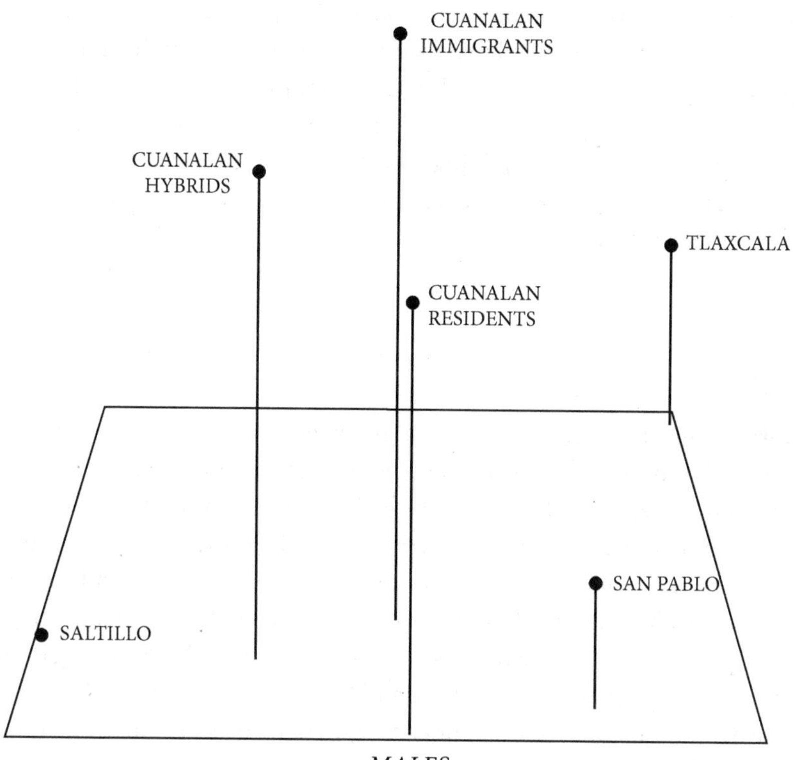

Figure 3.4 Three-dimensional scattergram of male group means on the first three canonical axes. This plot pictures the optimum separation of groups along each axis.

Source: Reproduced from Crawford, M. H. (Ed.), *The Tlaxcaltecans: Prehistory, Demography, Morphology and Genetics*. © 1976, The Author.

characterization of dermal ridges, patterns, and palmar lines were first proposed by Galton (1892), improved by Henry (1900), and further refined by Cummins and Midlo (1943). These methods for characterizing dermal variation between populations on a worldwide basis were initially applied by Cummins and Midlo (1943). They included some of the earliest investigations of dermatoglyphic variation in Mexican indigenous populations (Cumins, 1930). Finger and palm prints were collected from the Valley of Tlaxcala, Cuanalan, and Saltillo in 1969, 1972, and 1974.

A total of 910 participants provided dermal prints (187 from San Pablo, 120 from Tlaxcala, 348 from Cuanalan, and 255 from Saltillo). Males and females were analyzed separately because of the differential number of ridges associated with size of hand and gender. Comparisons of the four male Tlaxcaltecan groups revealed that the Saltillo sample (containing the highest Spanish ancestry) was the most different from the other three groups. The Cuanalan sample most closely resembles the Native American village of San Pablo del Monte, in the adjoining valley. The heterogeneity of the dermal traits, numbers of ridges, frequencies of patterns on fingers, and palm lines complicate genetic analyses.

Dentition

Morphological variation of dentition has been widely utilized in the characterization of hominin evolution. Teeth, because of their hardness, tend to be better preserved than other skeletal remains of the body. Dental casts have been utilized for the characterization of dental variation in contemporary human populations. Studies of morphological variation based on dentition utilize either odontometrics (measurements of size and shape) or the presence of discrete traits, such as cusp of Carabelli, shoveling of the incisors, and cusp numbers (Baume and Crawford, 1978, 1980). These discrete dental traits were scored for each individual using standard methods developed by Dahlberg (1956) and applied by graduate research assistant Robert Baume. In this research program on the transplanted Tlaxcaltecan populations, odontometrics and incidence of discrete traits were used to measure population affinities. However, these analyses were complicated by varying degrees of admixture and smaller sample sizes because there was less enthusiasm by participants in the process of making dental impressions. In addition, only adult dentitions were used and the sample sizes were further reduced by absence of the second and third molars occurring in 30% of the sample (O'Rourke and Crawford, 1976). Dental measurements (odontometrics) provided a reliable but conservative reflection of population history. Discriminant function analysis based on odontometrics shows significant differences between all populations but Tlaxcala City and San Pablo del Monte, the two populations from the same valley of origin. The genetic distances between the Mestizo populations are much smaller than between the Native American community and the Mestizo groups.

Paul Lin (1976) applied an analytical method, factor analysis, to Tlaxcaltecan odontometrics. He noted microevolutionary divergences in the underlying factors of the dentition for Cuanalan, San Pablo del Monte, and the City of Tlaxcala. The intergroup differences were evaluated based on the concordance in loading patterns between the metameres (one of a series of homologous parts) and between the members of a morphological tooth group. This method relies on a visual approach and requires a quantitative measure of agreement in factor structures. He also observed a general concordance between the morphological tooth classes and the discrete units defined by the distribution of factor loadings.

Discrete dental traits characterized morphological variation among the transplanted populations and the sole Native American community from their Valley of Tlaxcala (O'Rourke and Crawford, 1976; Baume and Crawford, 1978). Interpopulation genetic distances based on discrete traits revealed that Cuanalan and San Pablo are the most divergent groups, while the City of Tlaxcala and San Pablo del Monte clustered together. In these analyses, it appears that populations sharing similar environments (such as residents in the same Valley) exhibit similar morphology. Factor analyses of three of the populations (San Pablo, City of Tlaxcala, and Cuanalan) were suggestive of morphological divergences in the underlying factors of dentitions. Concise assessments of population affinities were not possible because of different levels of gene flow from Spanish and African groups into the transplanted populations. Reproductive isolation in both Tlaxcaltecan transplants was lost approximately two generations ago with the influx of immigrants (Crawford et al., 1976). The dental casts used in this study are available for possible reanalysis at the Crawford Collection, Maxwell Museum, Santa Fe, New Mexico.

Conclusion of Tlaxcala Project

This Tlaxcaltecan research program demonstrated that populations undergoing fission and relocation for 300+ years differentiate genetically and morphologically (Crawford, 1980). However, the genetic makeup of the original ancestral populations and the degree of gene flow after population transplantation complicates the interpretation of exact measures of differentiation. Admixture measures based on standard markers gave an almost identical estimate of ancestry as did the HLA system assessed by molecular genetic methods. This study also provides some insight into the underlying genetic bases of dental, anthropometric, and dermatoglyphic traits.

4

Origins of the Irish Travelers (Tinkers)

Introduction

As a graduate student, while working with Derek Roberts on an analysis of the genetics of the populations living on a small, isolated island in the Atlantic (Tristan da Cunha), I became interested in detecting genetic drift in human populations. A few years later, I learned from Eileen Kane (a former colleague at the University of Pittsburgh) that the southern coast of Ireland contained an assortment of small islands, inhabited by miniscule populations. In 1970, I contacted the Dublin-based Medico-Social Research Board (MSRB) of Ireland and inquired about the possibility of collaborative documentation of genetic drift on an island off the southern coast of Ireland. I received an enthusiastic response from the director of the Institute, Dr. Geoffrey Dean, who sent two physicians to an island to explore the possibility of research and the availability of health clinics and facilities to accommodate a collaborative team of researchers. The physicians inquired about the presence of specific facilities and the community willingness to participate in a research program. I was given the impression from MSRB that all arrangements had been made and then I applied to the Wenner-Gren Foundation for Anthropological Research for financial support for this project.

Participants

After being informed by Wenner-Gren Foundation that the project (grant number 2647) was funded, I solicited assistance from a cultural anthropologist, Dr. George Gmelch (University of San Francisco), to help reconstruct historical and demographic data as background and interpret the genetic characterization of the population. In addition, a dentist from Kansas City, Dr. Ted Rebich, agreed to provide dental care to the Irish population. In the summer of 1970, the three of us arrived in Dublin and met with Dr. Geoffrey Dean, director of MSRB of Ireland, to discover that the physicians that Dean

In Search of Human Evolution. Michael H. Crawford, Oxford University Press. © Oxford University Press 2024.
DOI: 10.1093/9780197679432.003.0004

had sent to the island managed to antagonize the island community. They had asked questions in an inappropriate manner concerning the availability of housing, refrigeration, and centrifuges in the clinics on the island and received a response of "Who do the Americans think we are?" "We are civilized and have all the available equipment!" The inhabitants found this exploration of the availability of facilities as condescending and refused to participate in any study associated with MSRB and any US researchers. Thus, on our arrival in Dublin, with a grant in hand, we did not have a viable study. While in Ireland, however, I observed the presence of Tinkers (also known as Travelers or Itinerants), who traversed the countryside in horse-drawn, colorfully painted barrel-top wagons with women and children begging on the streets of the city. Given their unique lifestyle, I inquired as to whether these travelers (Tinkers) had Romany Gypsy ancestry or were of Irish origin—and was told that no one really knows. They spoke a unique dialect or distinct argot, Gammon or Cant, which is unintelligible to the settled Irish populations.

Theories of Itinerant Origins

MacMahon (1971) summarized the major theories of the ethno-genesis of the Itinerants of Ireland: (1) The Itinerants are descendants of the outcasts who lived beyond "the circle of the ancient body of common law of prehistoric Ireland." (2) They are descendants of native chieftains, and their families were dispossessed by the successive English plantations. In support of this theory, McMahon claimed that the Tinkers of today bear the names of some of the so-called noblest Irish clans associated with territories inhabited by their chieftain ancestors. (3) The Itinerants are a result of admixture between the Irish and Romany Gypsies. (4) The Itinerants are displaced peasants and laborers driven from their lands and occupations by economic upheavals, famines, and conflict. Cromwell's conquest of Ireland (1649–1650) was followed by the Act of Settlement of 1652, passed by the Parliament of England. This Act imposed penalties, including death and land confiscation, against participants in the Irish Rebellion of 1641. This Act resulted in the seizure of Irish lands and forced some of the Irish population to assume an itinerant lifestyle.

The Tinkers assumed a traveler's lifestyle and were distributed geographically throughout the counties of Ireland (Bohn and Gmelch, 1976). However, the Irish government in the Report of the Commission on Itineracy wanted to settle the travelers in housing projects of Dublin (Government of Ireland, 1963). The majority of these travelers refused to settle in government housing and continued traveling their regular circuits, usually within the confines of one or two counties. Tinsmithing was one of their primary occupations, but some Travelers (blockers) specialized in trading and selling horses (usually in regional fairs), swept chimneys, peddled small wares, and performed various odd jobs in rural communities. With the decline in their traditional occupations, Traveler men turned to collecting scrap metal while women and children begged at specific street corners on the streets of Dublin.

By the late 1960s, a substantial majority of Ireland's Traveler population had migrated from the countryside to urban government settlements. In 1971, 248 of Ireland's 1,302 Traveler families were living on the outskirts of Dublin (Crawford and Gmelch, 1974). In response to a large number of migrants moving from rural to urban locations, the Irish government constructed housing for these itinerants. The Irish government built 70 serviced campsites specifically for itinerant families.

Currently, approximately 300 families are living on government sites. Some of these families are permanently settled in government housing while others remain at these sites a few months each year before resuming a traveling lifestyle (see Figure 4.1).

In 1970, approximately 300 families were residing in government housing in several sites surrounding Dublin. Some Travelers settled permanently, while the majority remained in government housing for a few months before hitting the road again. Approximately 800 families continued living along the roadside in their horse-drawn wagons. According to government censuses, the size of the Traveler population of Ireland was increasing exponentially.

The reason for this massive population increase is that the Irish Travelers exhibited one of the highest fertility rates recorded in contemporary human populations. The mean number of children per prolific woman 40 years of age or older is 10.43 with a variance of 22.39 (Crawford and Gmelch, 1974). The mean number of children per prolific woman in the total Traveler population is 7.94 with a variance of 2.68. This exceptionally high fertility is explained by the long reproductive careers of the females, who are married on average at 16 years of age and continue reproducing until 45 years of age. There appears to be significant differences in the achieved reproduction rates

Figure 4.1 Photograph of a Traveler couple posing in front of their horse-drawn wagon.
Source: Photo courtesy of George Gmelch.

of the recently settled Travelers from those following the more traditional lifestyle. The mean number of children in the government settlements is 9.73, while the Travelers who maintain a more mobile existence have on average 5.70 children. This fertility differential is most likely due to an earlier age of marriage, availability of potential mates in settlements, and a reduction in infant mortality. The settled Travelers have better access to health clinics that provide free medical care.

Methodology

During the summer of 1970, genealogical, demographic information and blood specimens were collected from 119 Irish Traveler participants in health clinics set up in the settlements and along the roadside of rural Ireland (Crawford and Gmelch, 1974; Crawford, 1975). Three governmental settlements on the outskirts of Dublin, Labre Park, Finglas, and Rathfarnam were the primary sites of investigation of the settled communities. Dental care was provided by a team dentist from Kansas City (Dr. Ted Rebich) and a certified dentist from Dublin (Dr. Liam Convery). This arrangement was necessitated because the Irish Dental Association requires local certification to practice dentistry. Ted Rebich, uncertified in Ireland, could not carry out any major clinical procedures without appropriate Irish supervision. The Irish Heart Association kindly provided 95 blood specimens from unrelated individuals drawn during their survey of the town of Kilkenny. For comparative purposes, allelic frequencies of genetic markers in other populations of Ireland and British Gypsies were compiled from the published literature (Crawford, 1975).

Results

Data on the frequencies of 33 alleles from 12 red blood cell loci were utilized for the genetic characterization of the Traveler population and compared with Eurasian populations, Romany Gypsy groups, and populations of Punjab, India (North et al., 2000). Table 4.1 lists the genetic distances among eight populations.

Genetic Affinities

Although standard genetic markers (14 alleles and chromosomal segments) are often less informative than DNA-based comparisons, the Travelers are closest genetically to average frequencies for Ireland and the Irish town of Kilkenny. However, Travelers are significantly different from the East Indian populations that have been proposed as ancestors of the Romany Gypsies of Europe. Despite the shared lifestyle with European Gypsies, Travelers differ genetically from Hungarian Gypsies.

Table 4.1 Genetic Distances among Travelers, European Populations, and Indian Populations Based on Sanghvi's X^2

	Travelers	Kilkenny	Ireland	England	Iceland	Hung. Gypsies	Hungary	Punjab
Travelers	0							
Kilkenny	1.150	0						
Ireland	1.446	0.674	0					
England	2.312	0.851	0.457	0				
Iceland	3.949	1.770	1.338	0.798	0			
Hung. Gypsies	3.269	1.840	4.574	4.200	5.341	0		
Hungary	1.823	2.479	2.109	1.702	2.823	0.658	0	
Punjab	3.088	3.214	6.007	6.487	8.231	0.868	2.144	0

Source: Reproduced from Crawford, M. H., "Genetic Affinities and Origin of the Irish Tinkers," in *Biosocial Interrelations in Population Adaptations*, ed. E. Watts, F. Johnston, and G. W. Lasker, pp. 93–103. © 1975, Mouton Press, The Hague.

Conclusions

Two primary hypotheses were posed in regard to the origins of the Irish Travelers:

1. Travelers are genetically related to the Roma Gypsies since they share a similar nomadic lifestyle and exhibit some cultural similarities with the Roma populations of Europe (Crawford, 1975; North et al., 2000).
2. Travelers are of Irish origin but have differentiated genetically from settled Irish populations through stochastic processes, such as genetic drift (Relethford and Crawford, 2013). Genetic distances based on frequencies of genetic markers of the blood reveal a close genetic affinity of the Travelers to the settled populations of Ireland (Crawford, 1975). However, we found that the Travelers are genetically distinct from a number of Irish populations. This drift hypothesis was tested by Relethford and Crawford (2013), comparing genetic distances of the Travelers to four geographic regions of Ireland. After adjustment for geography through distances, the genetic distances were compared, using the method developed by Relethford. The unadjusted distances revealed the genetic distinctiveness of the Travelers, who appear to be equidistant from other Irish populations. According to a mathematical model used in our study, the observed genetic differences reflect the action of genetic drift, since there is no evidence for external gene flow (Relethford and Crawford, 2013). This study of the Irish Travelers (Itinerants) demonstrated the action of genetic drift (stochastic processes) on small, isolated populations of Ireland traveling the roads of the rural countryside.

Contrary to MacMahon's (1971) hypothesis that the Travelers were forced to take to the roads by the famines of the 1880s, the historical accounts suggest that the Traveler population predated the potato blights and the agricultural plantations of Cromwell. The literature of both Ireland and England contains numerous references to "Tinkers" (now preferentially referred to as Travelers or Itinerants). For example, in the late 1500s, Shakespeare referred to the so-called Tinkers in several of his plays, most notably in *Henry IV*. In the early 1500s, legislation directed against the Tinkers was passed in Ireland. It appears that a number of historical events, such as the closing of monasteries due to Norman invasion, the potato famine, and agricultural droughts, forced the Travelers to take to the road and assume an itinerant lifestyle. The combination of famines, droughts, and repression by the British occupation thus created the current Irish Traveler (Tinker) society (North et al., 2000).

Travelers have a complex history with additions to the population being a function of various social and historical disruptions in Ireland. The potato blight, closing of the monasteries, social isolation, and the action of genetic drift all contributed to the genetic differentiation of the Travelers from the surrounding Irish settlements.

5
Black Caribs (Garifuna) of Central America
An Evolutionary Success Story

Introduction

In 1975, I was invited by Professor Thomas Weaver, colleague, friend, and former racquetball partner from the University of Pittsburgh, to give a presentation in a symposium that he organized on causes of migration at the Applied Anthropology Congress in Amsterdam, the Netherlands. During this meeting, Weaver attended a reception for the officers of the Applied Anthropology Association, and he took me along. At that reception, I met Nancie Gonzalez, an outgoing officer of the organization. She was interested in the genetics of the Black Carib population of Livingston, Guatemala, and wondered if their ancestry and gene flow (admixture) could be reconstructed using genetic markers. I recounted some of our past research in Tlaxcala and explained that if the population agreed to participate and provide blood specimens, then we could estimate their African and Native American ancestry using an assortment of genetic markers, such as blood groups, protein variants, and immunoglobulins. Being highly interested in a genetic approach, Nancie Gonzalez invited me to accompany her to Livingston, Guatemala, where she planned to continue conducting field investigations that summer. After the Congress in Amsterdam, I went back to Lawrence and hustled some financial support from the University of Kansas to cover the costs of fieldwork in Livingston. The primary focus of these investigations was to estimate the Carib Native American contribution to the formation of the Black Carib gene pool (Crawford, 1983, 1984).

I was also able to convince one of my former Ph.D. students, Dr. Paul Lin, to accompany me to Livingston and to examine morphological variation in the admixed population using anthropometric measurements. During World War II, Paul Lin had survived the Japanese occupation of Taiwan and

In Search of Human Evolution. Michael H. Crawford, Oxford University Press. © Oxford University Press 2024.
DOI: 10.1093/9780197679432.003.0005

was forced (as a teenager) to assist in the repair of the runway in Taipei, while being strafed by US fighter planes. During this time, Paul learned Japanese and later enjoyed needling me by addressing me as "anjin san." In the 1980 TV miniseries *Shogun*, a British navigator, Jon Blackthorne (played by Richard Chamberlain), was addressed as "anjin san" (boss) in feudal Japan. This TV program, based on James Clavell's 1975 novel, popularized this term among TV viewers of the United States.

Participants

That summer, Paul Lin and I flew to Guatemala City and were met at the airport by Dr. Reynaldo Mattorell, one of my colleagues from the University of Washington. We stayed in his apartment in Guatemala City, and he kindly provided a tour of Antigua, a city in the Central Highlands of Guatemala. During the 18th century, this city of Antigua was partially destroyed by a horrendous earthquake. Such earthquakes are frequent in Guatemala, because the country lies on a major fault zone that cuts across Guatemala and forms a tectonic boundary between the Caribbean and North American plates. Antigua's colonial architecture had been destroyed by severe earthquakes, and only remnants of some churches and administrative buildings escaped destruction and have survived to this time.

During our stay in Guatemala City, Reynaldo Mattorell took Paul and me out for a drink to an impressive mansion. We were served drinks but noticed a series of Asian businessmen pairing off with attractive women and disappearing from view. Several of these women approached us and offered their services. At this point, it finally dawned on both of us that Rey had taken us to a high-class Guatemalan brothel. We thanked these ladies profusely for their attention and kind offers but beat a rapid retreat from the premises.

The next day Paul Lin and I took a bus from Guatemala City to Puerto Barrios on the Caribbean coast of Guatemala, 297 kilometers northeast and the terminus of highway CA9. Puerto Barrios is the main port to the Caribbean seaport and a point of departure to Livingston. At that time, Puerto Barrios, consisting of 17,000 inhabitants, was a fascinating mix of Afro-Guatemalans, Maya, and Afro-Jamaicans. We boarded a boat in Puerto Barrios, which took us to Livingston, Guatemala, where Nancie Gonzalez met us and introduced us to the Garifuna population.

History of the Black Carib Population

Archaeological evidence from approximately 8,000 years ago exists for the initial settlement of the Lesser Antilles. The original home of the Black Caribs, St. Vincent Island, was first colonized by Native Americans from the Orinoco River Basin of South America (Fitzpatrick, 2015). There is some controversy as to whether the earliest settlers of this region were Arawak speakers or Caribs from Central America. The initial settlement of the Lesser Antilles followed a Carib invasion. European contact occurred in 1492. By 1503, the Spanish Crown had claimed St. Vincent Island. During the Atlantic Slave Trade, more than two million African slaves were forcibly relocated from the west coast of Africa to the Antilles (Davis and Knecht, 2010). By 1676, there were 3,000 Black Caribs (a mixture of African and Carib) living on St. Vincent Island. The French initially colonized St. Vincent Island; however, this island was transferred to British control under the Treaty of Paris. Currently, Livingston has a population of 18,000 persons and has become a major tourist attraction. During the 1790s, warfare for control of St. Vincent Island ensued between the British and the Black Caribs (Taylor, 2012). The British army prevailed and in 1797 several thousand Black Caribs were deported to Baliceaux, a small island south of St. Vincent. However, this island had a scarcity of drinking water, and during their imprisonment at least 1,000 of the Black Caribs perished from disease, most likely yellow fever or typhus.

The Black Caribs originated on St. Vincent Island of the Lesser Antilles and were descendants of shipwrecked slaves and Native American inhabitants of the island. In the 17th century, some African slaves escaped from the slave center on the adjoining island, Barbados, resulting in a mixed Native American/African population (see Figure 5.1). The Africans interbred with the local indigenous populations to create the Garifuna (Black Caribs) of St. Vincent Island. In 1797, after rebellion against the British colonizing forces, a total of 2,026 Garifuna were forcibly deported by the British Navy, first to a small island off the coast of St. Vincent, Baliceaux (see Figure 5.2) followed by relocation to Roatan, in the Bay Islands. From Roatan, the Garifuna migrants were transported by a Spanish fleet to the coast of Honduras, near Trujillo, and from there spread rapidly into Guatemala, British Honduras (Belize), Honduras, and Nicaragua. The Garifuna expanded numerically from fewer than 2,000 founders in 1800 (residing in two communities in Honduras) to more than 100,000 persons in 54 small horticultural and fishing villages

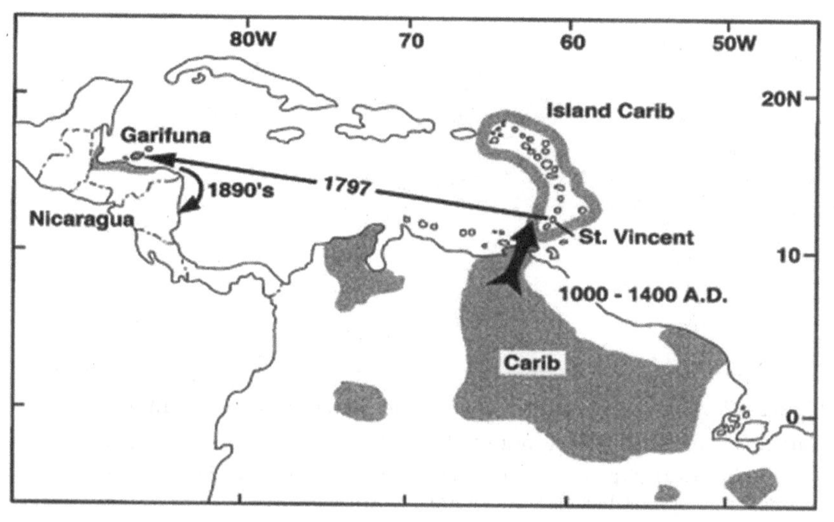

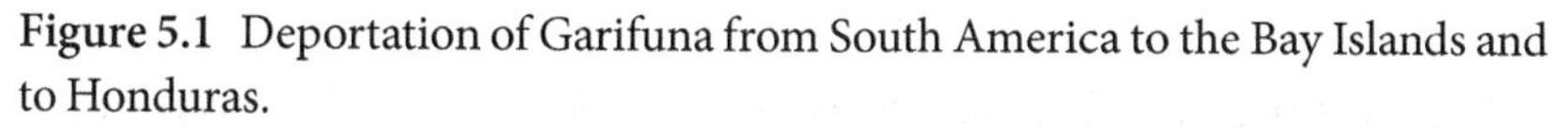

Figure 5.1 Deportation of Garifuna from South America to the Bay Islands and to Honduras.

Source: Reproduced with permission from Davidson, W. V., "The Garifuna in Central America," in *Current Developments in Anthropological Genetics*, ed. M. H. Crawford, © 1984, Springer.

located along the coast of Central America (Davidson, 1984) (see Figure 5.3). This unique numerical expansion of the Garifuna was followed by multiple fissions of the population as result of (1) favorable environmental conditions, (2) availability of food resources, and (3) exceptionally high fertility during the first few generations (Brennan's 1983 study of fertility among the Garifuna found on average each prolific woman had 10.9 children at the completion of her reproductive career), and (4) genetic adaptation to malaria, in the form of abnormal hemoglobin (sickle cell anemia and hemoglobin C), Duffy null, and G-6-PD deficiency; these were introduced into Central America primarily by African slaves working the fruit plantations. There is considerable evidence that the Coast of Central America had a relatively high incidence of malaria (Custodio and Huntsman, 1984). Judging from the absence of any genetic markers associated with resistance to malaria in Native Americans, the coastal populations of Central America experienced massive selection after African slaves brought Plasmodium organism (one cause of malaria) to the Americas.

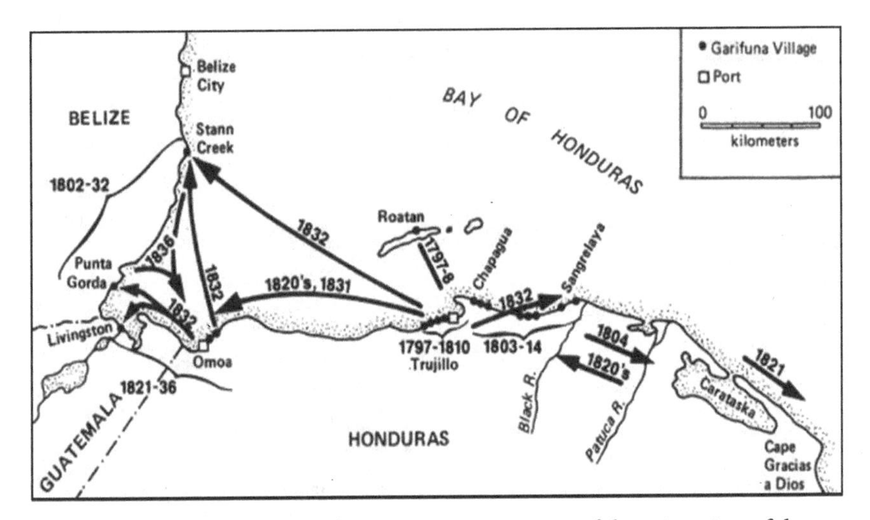

Figure 5.2 A chronological schematic representation of the migration of the Garifuna from Roatan to Honduras, Guatemala, Belize, and Nicaragua.
Source: Reproduced with permission from Davidson, W. V., "The Garifuna in Central America," in *Current Developments in Anthropological Genetics*, ed. M. H. Crawford, © 1984, Springer.

The Garifuna were forcibly relocated by the British forces of occupation, first south to Balliceaux Island and then to the Bay Islands, off the coast of Honduras. The Spanish authorities had political control of the Bay Islands and did not expect the deposition of several thousand Garifuna to their shores. Most of these migrants were not interested in residing in the Bay Islands and convinced the Spanish authorities to transport approximately 2,000 Garifuna to the coast of Central America. Because of the introduction of malaria to the coast of Honduras, Native Americans migrated inland while the coast of Belize, Honduras, and Guatemala was available for the rapid colonization by the Garifuna. Figure 5.2 reconstructs the rapid movement of the migrants from two Honduran villages (where the Garifuna originally settled) to 54 coastal communities distributed from Belize to Nicaragua.

This takeover of the coast of Central America by the Garifuna was facilitated by (1) exceptional fertility; (2) their resistance to malaria because of an African origin; and (3) a combination of horticultural and fishing culture that they brought from St. Vincent Island to the coast of Central America. The Garifuna population size increased dramatically from fewer than 2,000 individuals transported from Roatan in the Bay Islands to more than 100,000 distributed throughout Central America in the 1970s. In 1975, Livingston was a Black Carib (Garifuna) community of approximately 3,000 inhabitants located on the north Atlantic coast of Guatemala proximal to the Belizean border (Crawford et al., 2021). Currently, Livingston is a much larger population that attracts numerous tourists.

Figure 5.3 plots the broad coastal distribution of the Garifuna settlements, from Belize to Guatemala, Honduras, and Nicaragua (Davison, 1984). Following the transplantation of 2,000 Black Caribs from the Bay Islands to two villages in Honduras, the founding populations underwent fission into 54 settlements. The chronology of the migrations of Garifuna from Roatan to Honduras to Belize, Guatemala, and Nicaragua is shown in Figure 5.2, which also places the geographic locations of 54 Garifuna communities of Central America (see Figure 5.3).

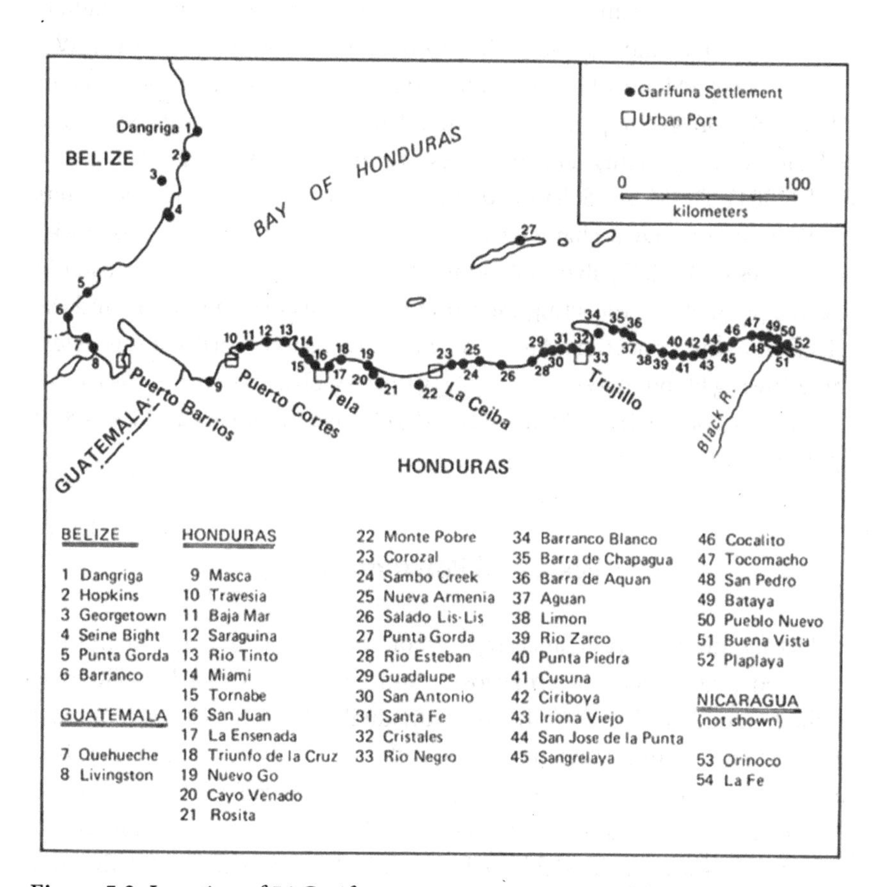

Figure 5.3 Location of 54 Garifuna communities scattered throughout the coast of Central America.

Source: Reproduced with permission from Davidson, W. V., "The Garifuna in Central America," in *Current Developments in Anthropological Genetics*, ed. M. H. Crawford, © 1984, Springer.

Field Research

Because of her longtime research experience in Livingston and political connections in Guatemala, Nancie Gonzalez obtained the necessary permissions from the Guatemalan government and local officials (Gonzalez, 1969). She also obtained individual verbal informed consent from 206 Garifuna who participated in the study, with blood samples drawn by venipuncture at the meeting hall of the local Catholic Church. During the summer of 1975, a research team formed consisting of Nancie Gonzalez, Paul Lin, and me. I collected blood specimens, finger and palm prints, anthropometric measurements, and genealogical/demographic data from inhabitants of Livingston, Guatemala. The blood samples, preserved in vacutainers containing ACD and packed in ice, were shipped for analysis to the War Memorial Blood Bank of Minneapolis, Minnesota, where laboratory director Chad Leyshon genotyped an assortment of erythrocyte antigens, serum, red blood cell proteins, and enzymes (Crawford et al., 1981; Schanfield et al., 1984). Dr. Moses Schanfield at the Red Cross National Headquarters in Washington, DC, characterized immunoglobulin (GMs and KMs) haplotypes in Black Carib populations. The GM haplotypes tend to cluster in specific populations reflecting their past disease history, migration, and the effects of natural selection. Ancestry of the Livingston, Garifuna was determined, using blood markers and GMs (gamma globulins), with 70%–75% of the genome coming from Africa, 22%–29% Native American, and 1%–2% of Spanish origin (Crawford, 1983).

Sickle Cell Polymorphism

Reflecting their African ancestry, 30 of the Garifuna from Livingston exhibited high genetic heterozygosity and genotype AS (sickle cell trait) at the hemoglobin locus. Only one individual exhibited sickle cell anemia (Hb SS); that is, two alleles for sickle cell. The Livingston population had an Hb S allelic frequency of 7%. In comparison, Firschein (1961), who had earlier sampled Garifuna for the presence of sickle cell genes, found an incidence of 24% with the sickle cell trait (Hb AS). Similarly, Custodio and Huntsman (1984) detected that the incidence of the sickle cell trait varied in Honduras from 15% to 19%. Duffy null allele-Fy (which provides resistance to Plasmodium vivax) ranges from 78% to 93% in Garifuna populations.

The genetic adaptation of the Garifuna to malaria added to their evolutionary success during their rapid expansion along the coast of Central America. Native Americans residing along the coast were at a selective disadvantage competing for regions parasitized by the Plasmodium falciparum and Plasmodium vivax organisms (Crawford et al., 2021). To avoid the consequences of malarial diseases, some of the indigenous Native American populations relocated from the coast of Central America into the highlands.

Firschein (1961) compared the reproductive rates of Garifuna women who were heterozygous for the sickle cell allele (Hb AS) with those homozygous for the normal adult hemoglobin (Hb AA). On average, Hb AS females produced 6.17 children during their reproductive careers compared to 4.25 livebirths for Hb AA women. Thus, in a malarial environment of coastal Belize, Hb AS females were 1.45 times more fertile than women with Hb AA. Firschein suggested that the Hb AA mothers experienced a greater number of pregnancy interruptions than the heterozygous mothers, resulting in loss of fetal lives with more males than females dying, reflected in the secondary sex ratio. However, Custodio and Huntsman (1984) found evidence that contradicted Firschein's findings. They observed that Hb AA females were more fertile than the carrier females with averages of 6.14 versus 5.0 livebirths for women 40 years of age and older. In addition, no statistically significant differences were observed in comparisons of the numbers of abortions and stillbirths between the Hb AA and Hb AS mothers. Research findings of Custodio and Huntsman (1984) contradict the earlier conclusion by Firschein (1984) that natural selection maintains the hemoglobin polymorphism through the greater fertility of the heterozygous females. In Africa, the hemoglobin locus is maintained at polymorphic levels through differential survivorship of children, particularly up to six years of age.

Garifuna of Belize

The tantalizing evolutionary results from the Livingston field investigations prompted me to expand the Garifuna research to a more intensive program in Belize. In the Livingston portion of the study, our data collection was limited because of the collective expertise of the three team members in cultural anthropology, genetics, and anthropometrics. In 1976, the Belize research program expanded to include the following methods and health services: (1) the Schroeder et al. (1973) and Schroeder (1974) micro-chromatographic method for identifying abnormal hemoglobin in the field without the availability of electricity (this permitted comparison of differential fertility of women with sickle cell versus those without any of the hemoglobinopathies); (2) application of anthropometric techniques to the assessment of growth and development of Garifuna children; (3) measurement of melanin concentration in the skin using reflectometry methods to assess admixture and gene flow; (4) assessment of morphological variation, based on dental morphology and dermatoglyphics; and (5) provision of medical care and reconstruction of individual medical histories by team physician Dr. David Hiebert. These additions to the Belize field research were supported financially by several universities through grants awarded to participating principal investigators.

This was a multifaceted project involving five institutions: University of Kansas, University of Kentucky, University at SUNY-Albany, Wichita State University, and the National Science Foundation (NSF). Since funds were allocated to individual institutions, human experimentation approvals had to be obtained separately through four different Institutional Review Boards (IRBs). Nancie Gonzalez obtained governmental permissions from Dr. CLB Rogers, Minister of Health, and Dr. K. Pike, Chief Medical Officer of Belize.

Participants

The research team of the Belize project during the summer of 1976 consisted of the following:

1. Elizabeth Murray, a graduate student in anthropology at the University of Kansas (KU), helped adjust a micro-chromatographic method, which could be adapted to the field because it was self-contained and did not require electricity to identify individuals with hemoglobinopathies. The construction of this apparatus based on publications by Schroeder and colleagues was under the guidance of KU biochemistry Professor Paul Kitos. Murray completed a Ph.D. in the Department of Genetics at KU.
2. Dennis O'Rourke, with previous field experience in Saltillo, Mexico, made amalgam impressions and poured dental casts from Garifuna willing to participate in this study (O'Rourke et al., 1984). O'Rourke completed his Ph.D. at KU in 1980, went on to a distinguished career at the University of Utah, and is currently a distinguished professor of anthropology at KU.

3. Because of her duties as program director of anthropology at NSF, Nancie Gonzalez joined the research team several weeks after the start of the project. During Gonzalez's absence, the cultural and genealogical information from Belizean Garifuna were collected by Loretta St. Louis, her former graduate student from Boston University (Gonzalez, 1984). After her term as program director at NSF, Gonzalez moved from Boston University and eventually was appointed the provost at the University of Maryland.
4. Highly experienced former Ph.D. student at KU, Professor Francis Lees, having conducted research in Cuanalan and Saltillo, Mexico, received a grant from SUNY-Albany (his first academic position) for research in Belize. He was accompanied by research assistant Pamela J. Byard, an M.A. student from SUNY Albany. They jointly utilized two reflectometers (British EEL and US Photo-volt) to measure melanin concentrations in research subjects and estimate admixture based on skin color in the Garifuna (Byard and Lees, 1981).
5. Professor Byard, having received her Ph.D. at KU, was appointed assistant professor and subsequently received tenure in the Department of Pediatrics at Case Western Reserve University. She left academia in the 1990s.
6. Professor Paul Lin (University of Chicago and Wichita State University) had collected finger and palm prints in Livingston and Belize and compared them to Spanish and African populations (Lin et al., 1984). He also published anthropometric measurements for Black Caribs of St. Vincent Island and Livingston (Lin, 1984).
7. Anthropometric measurements of the Garifuna volunteers were taken by Dr. Eugenie C. Scott, at that time an assistant professor at the University of Kentucky. She joined the research team because she wanted to switch research focus from skeletal biology to human biology of contemporary populations. Even though she standardized her measurements with those of Paul Lin, attempts at replication failed and her anthropometric data were never published. The only anthropometric data for Black Caribs came from Livingston and St. Vincent Island by Paul Lin. In 1976, Scott was a visiting assistant professor at KU, but her contract was not renewed. She did not get tenure at the University of Kentucky, nor at the Universities of Colorado and California at San Francisco. She eventually found her niche as director of the National Center for Science Education, which primarily supports the teaching of evolution in high schools (Scott, 2004).

8. The late Dr. David Hiebert, a physician and radiologist from Lawrence, Kansas, administered medical assistance and diagnoses to local physicians at the health clinics in Belize. He also served as the team physician on St. Lawrence Island, Alaska, in 1977.

Because of the strife among the researchers during the Belize field investigations, followed by litigation, I now prefer to limit field research to smaller teams with a single principal investigator, responsible for the all aspects of the research. Several of the researchers received funds from their own universities and exhibited varying degrees of training, competence, and research sophistication.

St. Vincent Island

I obtained permission to conduct field investigations on St. Vincent Island (an island in the Lesser Antilles) from the chief medical officer for the island, Dr. J. W. Kibukamusoke. He was Idi Amin Dadda's personal physician but had to leave Uganda after he fell out of favor with the dictator. Amin, a military officer, was known as the butcher of Uganda. He was president of Uganda from 1971 to 1979 when he was deposed. However, he managed to avoid justice by escaping to Saudi Arabia, where he spent the remainder of his life.

The Black Caribs (Garifuna) originated on St. Vincent Island as a cultural and biological amalgam between Native Americans (Arawak and Island Caribs) and African survivors of two sunken slave ships. A total of 2,026 Black Caribs were deported by the British in 1797 from St. Vincent Island to Baliceaux Island, and those who survived this confinement were transported to the Bay Islands and finally to the coast of Central America. Currently the descendants of the Garifuna who escaped British deportation (206 individuals) live in three villages (Fancy, Owia, and Sandy Bay) on the northeast slope of the volcano La Soufriere on St. Vincent Island. In 1979, three of the Black Carib villages from St. Vincent Island were sampled by a small research team consisting of Paul Lin, Michael Crawford, and Janis Hutchinson (Lin et al., 1984). For her doctoral dissertation, Janis Hutchinson focused on the relationship between the proportion of African ancestry and risk of hypertension. Usually African Americans experience a higher incidence of hypertension than subjects with only European ancestry (Rahman et al., 2008). Paul Lin measured anthropometrically a sample of the Black Caribs from the three communities. Several local nurses from the public health clinics served as phlebotomists. During the field research (April 1979), the volcano La Soufriere erupted, causing the relocation of some of the Garifuna from their villages of origin. An earlier devastating eruption in 1902 had killed 1,600 Caribs from the same three villages, namely Sandy Bay, Owia, and Fancy.

In the 1970s field seasons, we were limited to the use of protein variation, blood group markers, and immunoglobulin haplotypes. During the summers of 1975, 1976, and 1978, a total of 1,327 blood specimens were drawn from Garifuna and Creole volunteers from Belize, Guatemala, and St. Vincent Island (Crawford et al., 1984). Table 5.1 summarizes the number of participants in the initial anthropological genetic study of the Garifuna of Belize, Guatemala, and St. Vincent Island. The purpose of that study was to determine the evolutionary effects of forced migration on transplanted Garifuna populations (Crawford, 1983).

A summary of the blood genetic frequencies for Livingston, Stann Creek, Punta Gorda, Sandy Bay, Owia, and Limon is contained in a chapter of the volume on the Garifuna (Crawford, 1983). In addition to the allelic frequencies of an assortment of genetic markers, the proportion of African versus Native American admixture was calculated for six Garifuna population using a total of 26 alleles. See Table 5.1 for admixture estimates.

Table 5.1 Proportion of African versus Native American Admixture in Black Carib Populations of Guatemala, Belize, and St. Vincent Island

Carib Populations	African	Native American
Livingston, Guatemala	75%	25%
Stann Creek, Belize	67%	33%
Punta Gorda, Belize	70%	30%
Sandy Bay, St. Vincent Island	46%	53%
Owia, St Vincent	61%	39%
Limon, Honduras	64%	36%

Note: Since St. Vincent Island was an amalgam of Caribs/Arawak Native Americans, an average of gene frequencies of Arawak and Venezuelan Caribs was used to estimate ancestry.

Source: Reproduced with permission from Crawford, M. H., Dykes, D. D., Skradsky, K., and Polesky, H., "Blood Group, Serum Protein, and Red Cell Enzyme Polymorphisms, and Admixture among the Black Caribs and Creoles of Central America and the Caribbean," in *Current Developments in Anthropological Genetics*, ed. M. H. Crawford. © 1984, Springer.

Based on the calculation of Bernstein's admixture estimate of m, Livingston has the highest proportion of African genes, while Sandy Bay (St. Vincent Island) has the lowest, reflecting the history of their founding (Crawford, 1984). The founders of Sandy Bay, who escaped deportation to Central America, had the lowest African and highest Native American ancestry. In the use of genetic markers for the estimates of admixture, immunoglobulins (Gm and Km) and loci utilizing isoelectric focusing (IEF) were the most informative.

Quantitative Variation

Skin color variation (melanin concentration) was assessed among the Garifuna and Creoles of Bėlize, through skin reflectometry by Byard and Lees (1981). Creoles exhibit less melanin in the skin, and their melanin concentration was more variable than among the Garifuna (Byard and Lees, 1983). There was a concordance of admixture estimates based on skin color and blood genetic markers.

Black Carib and Creole populations were also compared morphologically using dental traits, odontometric traits, and odontological traits (Baume and Crawford, 1980; O'Rourke et al., 1984). Although Garifuna could not be distinguished from Creoles on the basis of discrete dental traits, odontometrics revealed significant differences between these two ethnic groups.

R-matrix analyses, based on finger and palm prints, reconstructed the genetic structure of the Garifuna populations (Lin et al., 1984). Significant correlations were observed between dermatoglyphics and geography—$r = 0.64$ for males and 0.35 for females—reflecting the patterns of gender influencing migration and population fission.

This research program on the Garifuna also examined several genetic epidemiological problems:

1. The relationship between African ancestry and essential hypertension was explored. In contrast to several earlier studies, Hutchinson and Crawford (1981) failed to demonstrate the expected greater risk of hypertension in individuals with high African ancestry. However, a statistical association was noted between body dimensions and blood pressure (Hutchinson et al., 1983).
2. In a collaborative study with NCI, field investigations on St. Vincent Island demonstrated a high incidence of antibodies against human T-cell leukemia/lymphoma virus that was similar to HTLV in American Blacks and identical to Japanese adult T-cell leukemia (ATL; Blattner et al., 1982).

DNA Results

During the 1980s and 1990s with the methodological developments in DNA extraction, characterization of DNA haplotypes and sequencing permitted much finer grained analyses of population differentiation and a more informative assessment of evolutionary change (Rubicz et al., 2007). Starting in 2005, a new series of follow-up investigations on Garifuna populations were initiated by faculty and graduate students from the Laboratory of Biological Anthropology (LBA) University of Kansas. See Table 5.1 for a summary of the samples collected in the Caribs of Dominica and Garifuna populations of Belize, Honduras, and Roatan in the Bay Islands. In addition, Noel Boaz (a physician, biological anthropologist, and faculty member of the Medical School at Dominica) and I characterized a population of Caribs from the Carib Reserve of Dominica. Dr. Boaz obtained signed informed consent, provided medical care, and collected blood through venipuncture from volunteers.

Resampling of St. Vincent and Belize

Christine Phillips-Krawczak (2012), at that time a doctoral candidate in human genetics at KU, conducted a follow-up study of the effects of migration on the Black Carib populations of Central America. She sampled the descendants of the original population of the Garifuna on St. Vincent Island and populations of Belize (Phillips-Krawczak, 2012). In addition, Kristine Beaty, a graduate student from the LBA, University of Kansas, accompanied a team of physicians from a medical school in Honduras and explored genetic variation and structure of three villages in Honduras plus the population of Roatan in the Bay Islands (Beaty, 2017).

Dominica

Dominica is the most northerly of the Windward Islands of the Lesser Antilles (see Figure 5.1). This island, first settled by Arawak-speaking Native Americans, was followed by the later expansion of Caribs from South America and absorption with gradual displacement of the Arawaks from the 1200s to the time of European contact. Due to warfare and newly introduced diseases, most of the Island Caribs did not survive contact with the European colonizers. Remnants of Carib-Arawak populations remain on Dominica. In the 1960 census, the total population of Dominica was 59,916 individuals, of whom 39,575 were of African origin, 19,606 were "mixed," and 251 had European ancestry. In 2013, the total population of the island of Dominica was 73,286 with 2.9% being of Carib origin.

In the 18th century, a Carib Reserve was established in the northeastern region of Dominica, in an inaccessible, steep, volcanic/forested region. The population size of the Carib Reserve, enumerated by the 1960 census, was 1,136 persons. The current estimate of the total Carib population of Dominica is 2,200, almost double the 1960 census size.

Results

In 2006, the Carib Reserve on Dominica was sampled by a research team consisting of Drs. Noel Boaz and Michael Crawford. Boaz, a biological anthropologist and physician trained in Dominica, was a faculty member of the Medical School on the island. He obtained written informed consent from 71 individuals who agreed to participate in the study. Kristin Beaty, research assistant at the LBA, following standard procedures, extracted DNA from the blood samples and analyzed them for genetic markers.

Mitochondrial DNA

DNA analyses included uniparental markers, mitochondria, and Y-chromosome. Mitochondria (small organelles in the cytoplasm of eukaryotic cells) originate in the mother's ova and are fertilized by sperm containing the Y-chromosome. Lynn Sagan (Margulis) (1967) suggested that the mitochondria descended from a symbiotic bacterium that found its way into an early ancestor of the eukaryotic cell. These ideas were labeled the endosymbiotic theory. The adaptive function of mitochondria is to provide cellular energy through respiration and oxidation. Human mitochondrial DNA are double-stranded (16,569 base pairs) circular DNA. It is highly conservative with no introns, and 90% is involved in coding. The coding region contains 37 exons (22 transfer RNAs, 13 proteins, two other RNAs). MtDNA is altered only by mutations and since it is uniparental, there is no recombination. The coding region of mtDNA contains a rapid mutation rate that is 5 to 10 times faster than nuclear DNA, at 0.07×10^{-6} substitutions per sites/year (Pakendorf and Stoneking, 2005). The hypervariable region (HVR) of the D-loop of mtDNA is located between regions that code for the t-RNA-proline and t-RNA-phenylalanine genes. The D-loop consists of three HVR regions—1, 2, and 3, each approximately 300 to 400 base pairs. The HVR sections of mtDNA display the greatest amount of variation and are usually first sequenced in most evolutionary studies.

Y-Chromosome Markers

The Y-chromosome is paternally inherited and can be used as an evolutionary marker for patrilineal descent. In humans, the Y-chromosome is approximately 60 million bases long, with 95% nonrecombining (NRY) Tilford et al., 2001. This chromosome is often characterized by single nucleotide polymorphisms (SNPs) and short tandem repeats (STRs). Five major haplogroups (R, I, J, G, and E) have been identified worldwide, with R1b being the most common marker. Y-chromosome STR mutation rate per marker generation is faster than the mutation rate in HVR of mtDNA.

Table 5.2 Sample sizes of Garifuna communities sampled in the study

Samples	Island/Country	Village	N
Previous Fieldwork	St. Vincent Island, 2004	Fancy	24
	Phillips-Krawczak (CPK)	Sandy Bay	34
		Owia	22
		Greggs	20
	Belize, CPK, 2005	Punta Gorda	50
		Barranco	27
	Dominica, MHC, 2005	Kalinago Reserve	71
Recent Fieldwork	Honduras Coast, 2014	Cristales	44
	K Beaty	Rio Negro	76
		Santa Fe	60
	Roatan, 2014	Punta Gorda	132
Y-STR data	Honduras Coast	Iriona	7
	Dr. Matamoros	Bajaamr	26
		Corozal	20

Source: Reproduced with permission from Crawford, M. H., "The Anthropological Genetics of the Black Caribs (Garifuna) of Central America," *Yearbook of Physical Anthropology* 26: 161–192. © 1983, Wiley.

The analyses of Garifuna genomes included mtDNA HVS I and II sequences (81–400 and 16050–16400) and RFLPs for subtypes, NRY SNPs (M9, DYS199, YAP, 92R7, and SRY), and eight STR loci. The majority of the mitochondrial DNA haplogroups—C1 at 9% A2 (55%)—were of Native American origin. The remaining mtDNA haplogroups were of African origin: L1b1, 4%; L1c, 6%; L2a, 4%; L2b, 2%; L3, 6%; and L3e, 9%.

Molecular studies of mtDNA in Dominica first began in 2006, with field investigations by M. H. Crawford and N. Boaz on the Kalinago Carib Reserve and analyzed by Phillips-Krawyczak, 2012 at the LBA. This was an early attempt to reconstruct the genetic structure of Carib ancestral populations. The Native American contribution to the currently admixed population of the Reserve is 67%, based largely upon the presence of Native American haplogroup C (57%) and A2 (10%). In contrast, Benn Torres et al (2015) found that the majority of individuals who they sampled on the Island of Dominica had African mitochondrial DNA haplogroups, with L being the most common. They reported that in Dominica only 28% of the lineages were of Native American, with a small founding population, high nucleotide diversity indicative of high level of admixture. Benn-Torres et al (2015) also reported, based on mtDNA, an extremely high African ancestry of 63%, for St. Vincent Island Black Caribs.

The study by members of the LBA reconstructed the genetic structure of Carib ancestral populations. The Native American contribution to the currently admixed population of the Dominica Reserve is 67%, based largely upon the presence of mtDNA haplogroups C (57%) and A2 (10%). In contrast, Benn Torres et al (2013) found that the majority of individuals who they sampled on Dominica exhibited African mtDNA haplogroups, with L being the most common. They reported that in Dominica only 28% of the lineages were of Native American, with a small founding population, high nucleotide diversity indicative of high level of admixture. Ben Torres et al. (2015) also reported, based on mtDNA, an extremely high African ancestry, 63%, for St. Vincent Island Black Caribs. Why the huge differences and conclusions in the two studies of the same islands? These studies indicate that admixture estimates are determined by who was sampled in the population, their geographic locations and what genetic markers are used. These discrepancies are due to sampling procedures: Torres study failed to characterize population substructure of the island populations. The LBA samples were focused on the presence of Native American subpopulations on Dominica (Carib Reserve) and St. Vincent (Carib villages on the slope of the volcano). Historical events play major roles in explaining island population structure and the distribution of marker genes.

Harvey et al. (1969) conducted the initial study on the genetics of the Caribs of Dominica. Blood samples from 126 volunteers were tested for an array of blood group antigens and a series of red cell enzymes and serum proteins. These samples resembled Native Americans with O blood group frequencies of 96%, Duffy (Fy) with a frequency of only 21%, and hemoglobin S at 2%. They concluded that the population had experienced some African gene flow but at low frequencies.

Chris Phillips-Krawczak genotyped the mtDNA haplotypes for her Ph.D. dissertation. The characterization of island populations is a function of sample sizes and geographic locations. St. Vincent Island and Dominica Caribs display considerable Native American ancestry with the considerable presence of A2 and C1 (see Figure 5.4). In contrast, the mtDNA haplotypes in Belize and Honduras primarily exhibit African mtDNA haplotypes L0–L3 with a miniscule incidence of Native American haplotypes—A2 and CI. It is likely the Black Carib populations expanded into Central America and admixed with Creoles.

Table 5.3 summarizes the frequencies of mtDNA haplotypes in populations of St. Vincent and Dominica and surprisingly shows vast differences in frequencies of the same populations. Specifically, the Torres group observed

Table 5.3 Admixture Estimates of Garifuna populations

Population	African	European	Native American
St. Vincent Island Caribs			
Sandy Bay	41.1%	16.7%	42.2
Owia	57.9	9.8	32.3
Belize Garifuna			
Stann Creek	79.9	17.1	7.4
Livingstone	70.0	1.0	29.0
Creoles			
Belize City	74.7	16.7	8.6

Note: CPK is Christine Phillips-Krawczak; MHC is Michael H. Crawford; and K Beaty is Kristine Beaty.

Source: Adapted with permission from Beaty, K. G., *Forced Migration and Population Expansion: The Genetic Story of the Garifuna*. Ph.D. diss., University of Kansas, 2017.

little Native American ancestry in both Dominica (28%) and St. Vincent Island (3.6%). In contrast, the LBA group found 60% Native American ancestry on Dominica and 43.4% on St. Vincent. The Black Caribs on St. Vincent Island are located in three villages on the northern slopes of the volcano. Samples of populations other than Sandy Bay, Owia, and Fancy yielded a Creole population. Similarly, samples obtained from Dominica in regions other than in the Reserve did not yield Black Carib frequencies but those of Creole populations.

Roatan

For her dissertation research, in 2016, Kristin Beaty joined a team from the Honduran Medical School led by Dr. Hernandez. The Honduran team sampled three communities in Honduras, where the original community from Roatan had been transplanted. This preliminary study was followed the following year by a sample from Roatan in the Bay Islands (Herrera-Paz and Mattamoros, 2010).

Figure 5.4 compares the frequencies of mtDNA haplotypes of Garifuna populations with those from Africa and the Americas. The two island Carib populations, St. Vincent and Dominica, cluster closely and reflect evolutionary history and migration. A more recent study of the population structure of Dominica using molecular markers measured genetic ancestry and population structure in rural populations of Dominica (Keith et al., 2021).

Conclusion

The molecular genetic data reveal that there was a reduction in genetic diversity due to a series of population bottlenecks (Beaty, 2017). This reduction

Table 5.4 Comparison of Admixture Estimates Based on mtDNA by Torres et al. and the KU Group for Dominica and St. Vincent

	Dominica		St. Vincent	
Parental Populations	LBA	Torres et al.	LBA	Torres et al.
Native American	60%	28%	43.4%	3.6%
African	40%	72%	56.7%	92.7%
European	0.0	0.0	0.0	1.7%

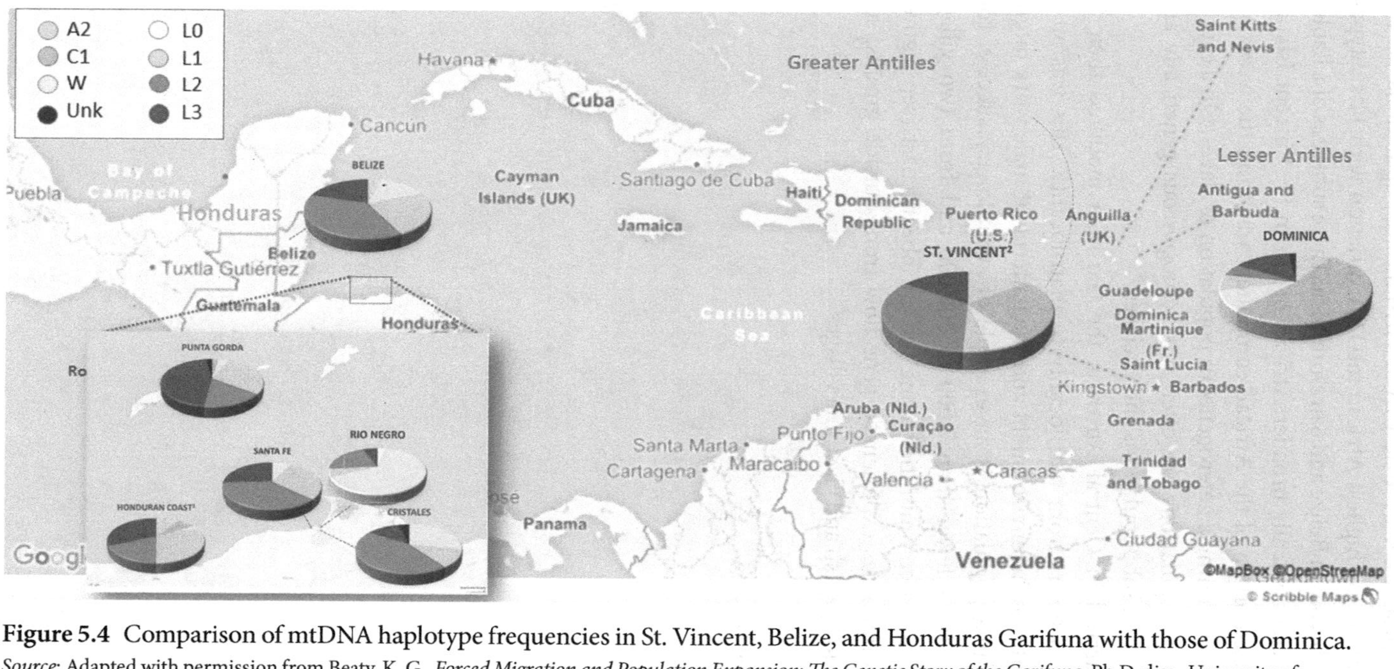

Figure 5.4 Comparison of mtDNA haplotype frequencies in St. Vincent, Belize, and Honduras Garifuna with those of Dominica.

Source: Adapted with permission from Beaty, K. G., *Forced Migration and Population Expansion: The Genetic Story of the Garifuna*. Ph.D. diss., University of Kansas, 2017.

took place despite the gene flow from surrounding populations. St. Vincent Island exhibits the highest African mtDNA variation with 11 subtypes of L versus 7 subtypes in Belize and 6 in Dominica. These subtypes of L haplotypes reflect a geographically broader African origin in St. Vincent Island. The hypothesized origin of the African component was from the wreckage of a slave ship containing a heterogeneous population of slaves. On a cautionary note, island populations are often not homogeneous genetically and are structured into subpopulations and enclaves.

After arrival in Central America, the Garifuna modified their family residence patterns from patrilocality to matrilocality. This was an adaptation to the social environment and the patterns of male work migration. The Garifuna population is an example of evolutionary success, roughly increasing from fewer than 2,000 to 300,000 and from two villages in Honduras to 54 communities along the Caribbean coast (see Figure 5.3). These communities experienced gene flow from outside groups, such as Creoles and Spanish and Garifuna villages.

6

Biological Aging and Population Structure of Midwest Mennonites

And so from hour to hour we ripe and ripe
And then from hour to hour we rot and rot,
And thereby hangs a tale

—William Shakespeare, *As You Like It*

This research program on the genetics of biological aging had its roots in 1975, when I received a telephone call from Professor Solomon Katz, a colleague from the University of Pennsylvania. Knowing that I was a trained geneticist who was fluent in Russian, Katz wanted to know if I would be interested in serving as a consultant to a study of longevity among the so-called yogurt-eating *dolgozhitili* (long-living) of Abkhasia and Georgia of the USSR. At that time, the literature was replete with examples of claimed longevity on a worldwide basis. Alexander Leaf championed extreme longevity in Georgia, Russia; in Hunza, Pakistan; and in the Karakoran Mountains (Leaf, 1973). Earlier, George Coggeshall described a longevity haven in Vilcabamba, Ecuador. However, Mazess and Forman (1979) on the bases of demographic data challenged the claimed extreme longevity of the inhabitants of this Ecuadoran valley. The 1970 census of Abkhasia listed 12% of the population as centenarians. In the 1970s, there was considerable controversy on the existence of hidden valleys with exceptional longevity attributed to genetics or environmental factors such as low-calorie diets, consumption of yogurt, and physical activity.

The Research Institute for the Study of Man (RISM) at Columbia University was planning to develop a research program in regions of the Soviet Union where there were claims of exceptional longevity and an apparent slowdown of the aging process. While in their 90s and 100s, Abkhasian men were reputed to be so spry as to be able to dance, to drink wine and vodka, and at such advanced ages to "chase women" (see Figure 6.1). Together with RISM researchers, Drs. Vera Rubin (Director of RISM) and Sula Benet (author of

In Search of Human Evolution. Michael H. Crawford, Oxford University Press. © Oxford University Press 2024.
DOI: 10.1093/9780197679432.003.0006

Figure 6.1 Photograph of three *dolgozhitili* (long living) from Abkhasia, USSR, all claiming to be over 90 years of age.

popular volume, *How to Live to Be 100*; Benet, 1976), I helped outline a bio-cultural research program for a study of longevity among the populations of Abkhasia and Georgia. The International Research and Exchange Board (IREX) sent the three of us (Sula Benet, Vera Rubin, and me) to Moscow, Kiev, Tbilisi, and Sukhumi to develop an international collaborative program on bio-cultural dimensions of the aging process. An agreement for a joint US-USSR research program was signed by the researchers from the Institute of Anthropology in Moscow and US researchers from RISM. We first applied for financial support from the National Institute of Aging (NIA) for a pilot study to standardize methodologies for the assessment of biological aging. The pilot study was funded (by NIA) through RISM and preliminary fieldwork began in Abkhasia.

A pilot study in Abkhasia on the *dolgozhitili*, using an anamnestic method (recalling to mind), revealed that there was significant exaggeration of the claimed ages. Each purportedly longevous person was instructed to identify the main stages of his or her life history and to connect them with specific dated events—such as the great blizzard of 1911. The purported ages were verified in only 38% from the 115 *dolgozhitili* interviewed. When the mortality curves were corrected for age exaggeration and differential emigration of the young added to the equation, it became evident that the survivorship patterns in Abkhasia did not differ significantly from those observed in Miami, Florida, populated by retirees from northern cities. This exaggeration of age was one strategy Abkhasian males used to avoid Russian military service. In addition, the elderly of Georgia and Abkhasia are highly revered and specially honored by their respective societies. This same pattern of age exaggeration occurred in another hidden valley of longevity, Vilcabamba, Ecuador (Mazess and Forman, 1979).

Midwestern Mennonites

Since our pilot study helped drain Ponce de Leon's imaginary fountain of youth, I decided to refocus my research efforts from the populations of the USSR to the nearby Mennonites of Kansas and Nebraska (Crawford and Rogers, 1982). The Mennonites exhibit greater life expectancy and longevity than the surrounding farming communities of Kansas (82 years of age for males and 86 for females) because they do not smoke or drink alcohol, lead an active agricultural lifestyle, and have a closed social support system. In addition, the Mennonites of Kansas and Nebraska have exceptionally accurate birth and death records, have large extended families, and are enthusiastic about learning about their origins and the genetics of biological aging.

General Conference (Dyck, 1993) and Haldeman Mennonites reside primarily in Goessel and Meridian in central Kansas (see Figure 6.2). They constitute a well-defined religious isolate group with the following characteristics: (1) they have a unique immigration history; (2) they have extensive written genealogies; (3) they have complex, extended family relationships permitting the study of heritability and linkage of biological aging; (4) they have genetic, environmental, and cultural homogeneity, which decreases the background noise for the polygenic study of biological aging; and (5) they do not smoke or drink alcohol and thus have an average life expectancy greater than those in surrounding populations.

Mennonites are an Anabaptist group that participated in the European Reformation of the 16th century (Rogers, 1984). A number of sects arose in Switzerland, Germany, and the Netherlands, who believed in adult baptism, the separation of church and state, and pacifism. They followed charismatic leaders, such as (1) Menno Simons, a Dutch-North German branch (became known as the Mennonites) and (2) Jacob Hutter,

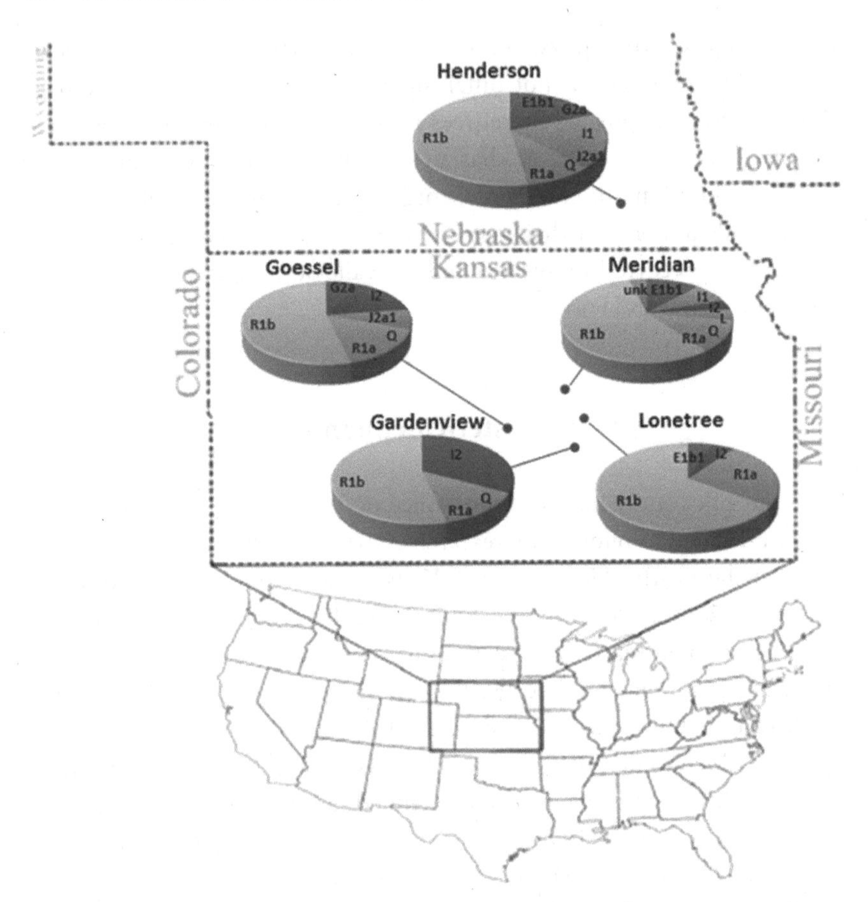

Figure 6.2 Map locating the six Mennonite communities from Kansas and Nebraska that participated in the study of genetics and biological aging.
Source: Adapted with permission from Beaty, K. G., Mosher, M. J., Crawford, M. H., and Melton, P., "Paternal Genetic Structure in Contemporary Mennonite Communities from the American Midwest," *Human Biology* 88(2): 95–108. © 2016, The Authors.

an Austrian branch of Anabaptists (known as Hutterites or Hutterian Bretheren). (3) Jacob Amman's followers were from Switzerland and were designated Amish or Amish Mennonites. Local authorities viewed the Anabaptists as a threat to the social order and began persecuting them. This resulted in the migration of the Anabaptist groups first moving to underdeveloped regions of Eastern Europe and the Americas. Dutch and German Mennonites migrated to Polish-controlled regions around Danzig (see Figure 6.3). In 1699, 18 families formed the Przechova church and relocated in 1821 to the Ukraine near the Molotschna River (Rogers

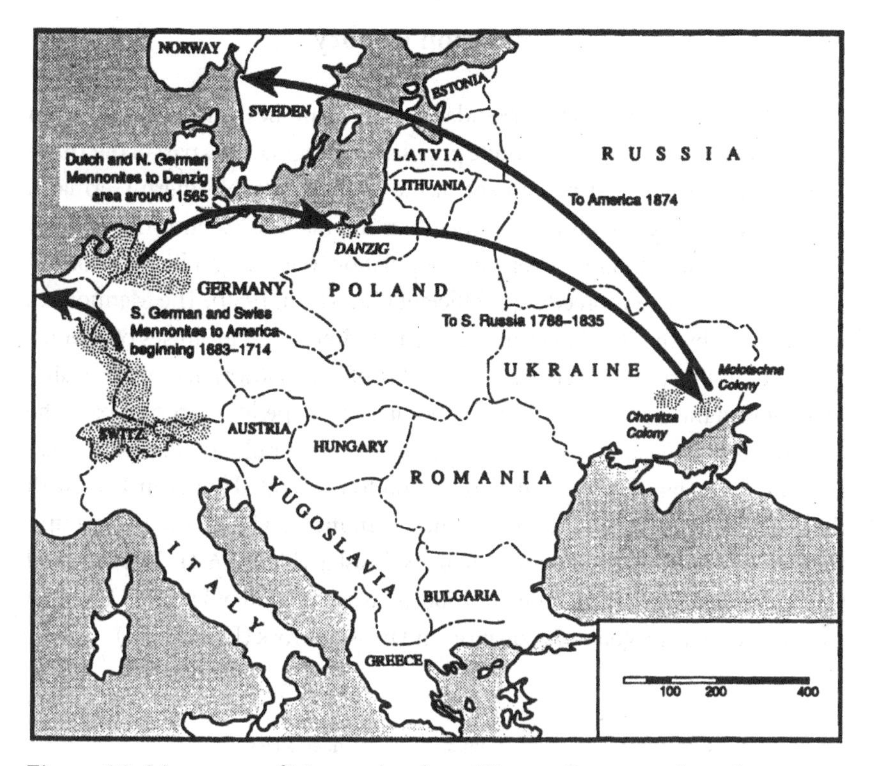

Figure 6.3 Movement of Mennonites from Western Europe to form the Molotchina colony in the Crimea.

Source: Reproduced from Crawford, M. H. (Ed.), *Different Seasons: Biological Aging Among the Mennonites of the Midwestern United States*, p. 39, fig. 4.9. © 2000, The Author.

and Rogers, 2000). This congregation adopted the name Alexanderwohl, in honor of the Russian czar (see Figure 6.4). Subsequent changes in economic conditions and shifts in Russian governmental polices concerning military exemptions prompted the Alexanderwohl Mennonites to purchase land from the Santa Fe railroad and migrate to the United States in 1874. Upon arrival in the United States, the Alexanderwohl community split into two separate congregations, one in Henderson, Nebraska, and one in Goessel, Kansas. A separate congregation (Church of God in Christ) of Mennonites, founded in Ohio in 1858 by John Holderman, was organized in Kansas. This heterogeneous community, consisting of Pennsylvania Dutch, German Mennonites, and Amish, formed the Meridian community in Kansas. This Holderman Mennonite community further split into Garden View and Lone Tree.

Population History

The Kansas Mennonites are descendants of 191 families who migrated from Europe in 1874. All of the Mennonite populations originated as part of the 16th–17th-century Anabaptist movement in the Netherlands, Northern Germany, and Switzerland (see Figure 6.4). They first migrated to Central Europe and then to Crimea, Ukraine, where the population was geographically isolated and exhibited a high inbreeding coefficient (Crawford, 1980). The Mennonites were enticed by the Russian czar to settle in Crimea with promises of exemption from military service and the availability of land. However, the succeeding czar revoked these privileges. The Mennonites sent expeditionary groups to the United States and purchased land from the Santa Fe railroad. Population fission was along familial lines after migration to the United States in 1874 with the General Conference Alexanderwohl community dividing into two groups, Goessel in Kansas and Henderson in Nebraska (Figure 6.4). The General Conference Mennonites, formed in 1860, is an association of US and Canadian conferences sharing goals of education and mission work (Dyck, 1993).

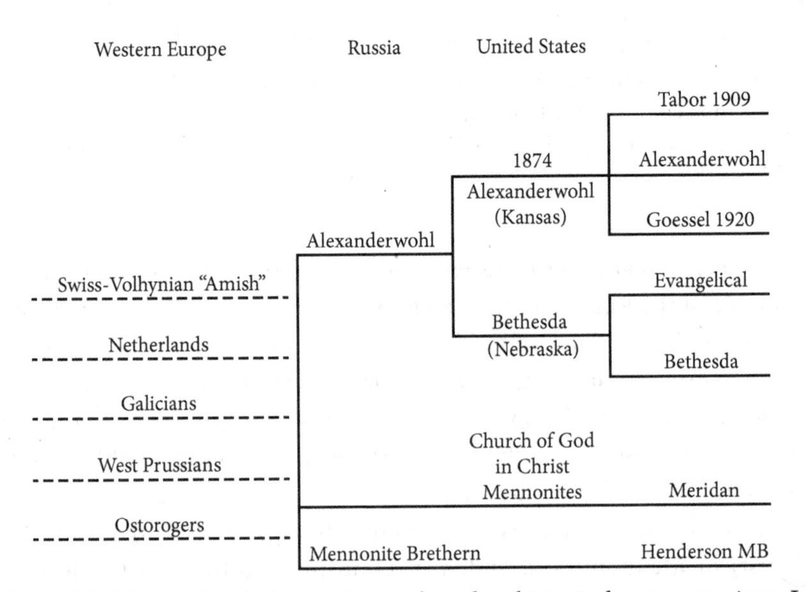

Figure 6.4 Mennonite phylogenetic tree, based on historical reconstructions. It contains the ethnic contributions from Western Europe to the Alexanderwohl congregations.

Source: Reproduced from Crawford, M. H. (Ed.), *Different Seasons: Biological Aging Among the Mennonites of the Midwestern United States*, p. 37, fig. 4.5. © 2000, The Author.

Meridian was the first Halderman community that we sampled in 1981. Lone Tree and Garden View Mennonites were offshoots from the original Meridian community, much more heterogeneous genetically, formed by the followers of charismatic leader Halderman. His converts included both Mennonites and Amish (see Figure 6.5). The Halderman offshoots, Lone Tree and Garden View, were sampled in 2004.

In 1979, the NIA (AG 01646) funded a three-year research program on the genetics of biological aging of a transplanted Mennonite community, who emigrated from Crimea to Kansas and Nebraska. The leaders of the Molotschna colony decided to emigrate from Crimea because land availability became scarce and the Russian government revoked their exemption from military service (Rogers and Rogers, 2000). The Mennonites of Gossel

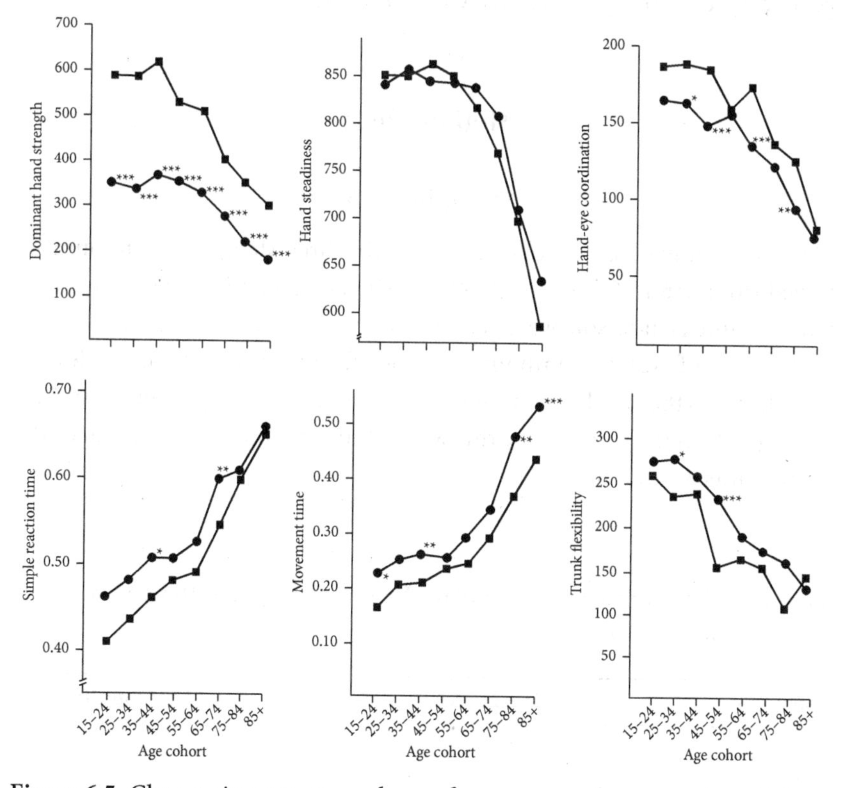

Figure 6.5 Changes in neuromuscular performance as a function of age. Values for males are shown as squares; females as circles.

Source: Adapted with permission from Devor, E. J., Crawford, M. H., and Osness, W., "Neuromuscular Performance in a Kansas Mennonite Community: Age and Sex Effects in Performance," *Human Biology* 57(2): 202, fig. 1. © 1985, The Authors.

and Henderson are the descendants of the 1,063 passengers who arrived in the United States in 1874. A total of 191 founding families were members of the original Alexanderwohl church from the Molotschna colony of Crimea, Russia. Meridian Mennonites have a more complex ancestry from the two Alexanderwohl communities with converts from several Mennonite and Amish communities. In addition to the study of genetics of biological aging, the transplanted Mennonite community underwent initial fission into two populations, one in Goessel, Kansas, and another in Henderson, Nebraska. The Goessel Mennonites planted winter wheat that they brought from the Crimea, while the Henderson Mennonites planted corn twice a year and became economically well-off. A genetic comparison of the two subdivisions of the original Alexanderwohl congregation provided an opportunity to assess how historical events (such as fission and familial founding) influence the genetic drift of small populations (Crawford, 2000).

Methodology

Sample Sizes

Maximum sample sizes of volunteers from the three Mennonite communities varied from 616 in Goessel and 549 for Henderson and 87 from Meridian. Goessel and Henderson were larger communities, and the percentage of participants of the total communities varied from 47% to 51% (see Table 6.1). Meridian was the smallest community but had the highest percentage of 54% participants in the study. Since this was a study of biological aging, all of the participants were adults.

Table 6.1 Maximum Sample Sizes of the Volunteers from the Three Mennonite Communities

Population	Sample Size (N)	% of Total Community
Goessel	616	47
Meridian	87	54
Henderson	549	51
Total	1,252	

Source: Reproduced from Crawford, M. H. (Ed.), *Different Seasons: Biological Aging among the Mennonites of the Midwestern United States*. © 2000, The Author.

Participants

Field investigations in Goessel and Meridian, Kansas, began during the winter of 1979–1980 (see Table 6.1). In 1979, the research team set up a clinic at the Alexanderwohl church meeting hall in Goessel, Kansas. This research constituted a partnership between the Mennonite communities and the University of Kansas.

Several Mennonites from the Goessel community were members of the research team, namely: two physicians—Drs. Peter Hiebert and his son and radiologist David Hiebert, cultural anthropologist Professor John Janzen, and psychologist Douglas Penner. This team also included researchers from Lawrence, the University of Kansas, and Cornell University:

1. Dr. Ralph Reed (internist MD from Lawrence)
2. Mennonite nurses from Central Kansas
3. Four University of Kansas (KU) exercise physiologists (Professor Wayne Osness and three of his graduate students)
4. Donald Stull, Jeff Longhofer, and Jerry Floerish (cultural anthropologists). Longhofer (1986) wrote his dissertation on the relationship between land, household and community.
5. KU sociologist Dr. Jill Quadagno
6. Graduate student from Cornell University, Joanne McGreevy (student of collaborating colleague, Dr. Jere Haas)
7. Social psychologist from the Department of Psychology, University of Kansas, Dr. William Bowerman
8. Five biological anthropologists also participated in the study, three faculty members—Tibor Koertvelyessy (visiting professor from Ohio University), Paul Lin (from Wichita State University), and Michael Crawford. Tibor Koertvelyessy tested the Mennonite participants for taste variation of PTC (Koertvelyessy et al., 1982). They did observe a low-level correlation between biological aging and loss of taste perception. However, this relationship was not due to the aging process of the taste buds but is most likely the result of smoking.
9. Two KU graduate research assistants from the Department of Anthropology, Laurine Oberdieck Rogers and Janis Hutchinson, played prominent roles in the organization of the study and data collection. The three physicians, through medical examinations and interviews, reconstructed the health histories of the participants, while the exercise

physiologists measured the neuromuscular, pulmonary function of all participating Mennonites. The cultural anthropologists collected genealogical and demographic data from the three communities.

Anthropometric measurements were individually taken by three members of the research team, namely Paul Lin, Laurine Oberdieck Rogers, and Joanne McGreevy (Cornell University). However, to minimize interobserver error, each investigator only measured the specific body dimensions in all of the research subjects. One researcher measured the head, another measured linear body dimensions, and a third focused on circumferential measures and weight.

Field investigations were limited to the winter months, when the Mennonite farmers were freed from time-consuming agricultural activities. Field investigations on such a massive scale were made possible by the superb administrative skills of research assistant, Laurine Oberdieck Rogers, who coordinated the scheduling, transportation, accommodations, and meals (prepared by the church members) for all researchers (Crawford and Rogers, 1982). Each researcher was assigned to his/her own station (worktable) plus an assistant who recorded the data/measurements. The clinics held at the meeting hall, adjacent to the church, were arranged into numbered "stations," so that each volunteer was met at the door and after giving informed consent was escorted to the appropriate and available station. In this way, there was a continuous flow of research subjects moving rapidly through the clinic.

Because of the size, diverse specialties of the research team, and the cooperation and interest of the Mennonite communities, a total of 1,252 Mennonites from three communities participated in this research (see Table 5.3). This sample represented large proportions of the total sizes of the communities; 47% of Goessel residents participated, 54% volunteered from the smaller, more heterogeneous Meridian community, and 51% of the Henderson community volunteered for the study. Since this research program focused on biological aging, only adults were included in the sample. The following neuromuscular performances were assessed: motor time (MT), simple reaction time (SRT), hand-eye coordination (HEC), hand steadiness time (HST), dominant grip strength (DGS), and trunk flexibility (TF). Professor Wayne Osness and his graduate students from the Exercise Physiology Department at KU also measured pulmonary functions (FVC and FEV1.0). Physicians assessed the health status of the Mennonites and measured heart rates (HR), systolic blood pressure (SBP), and diastolic blood pressure (DP) (Crawford, 2005). A total of 35 anthropometric traits were measured by three of the biological anthropologists, each measuring either the linear traits (which had the highest transmissibility), cranial measures, or circumferential traits (most affected by the environment). Blood specimens, drawn by nurses, were analyzed for 35 blood chemistry markers, ranging from total cholesterol to glucose levels to potassium levels (see Uttley and Crawford, 2000, for a complete listing of the blood chemistry panel). Lisa Martin focused her M.A. thesis on genetic/environmental interaction of thyroxine variation in Mennonite populations (Martin, 1993).

Neuromuscular and Pulmonary Traits

The neuromuscular and pulmonary traits measured in this study are all a function of the chronological age. As an individual ages, his/her pulmonary function diminishes, strength deteriorates, and reaction/motor times increase (see Figure 6.5). Morphologically, body mass index (BMI) and most circumferential measures increase with age (Devor et al., 1985). Only hand steadiness fails to show sex difference in mean performance levels. The strongest mean difference is on the dominant hand strength, in which average grip pressure in males is more than one and one half times greater than in females. However, females exhibit greater trunk flexibility than males (Devor, 2000). In addition, most of the blood chemistries varied as a function of age. A linear or curvilinear decline in neuromuscular and pulmonary performance was observed as a function of age beginning at 50 years and declining rapidly until death (see Figure 6.5).

Longevity as measured as a phenotype in years contains limited information on functional status. Mennonites live longer than non-Mennonite farmers in the surrounding regions with female life expectancy of 84 years of age, while male life expectancy is 82. The interesting question is which neuromuscular, pulmonary function, blood chemistries, and morphology are associated with greater longevity? We created a functional profile relative to an individual's peers. Chronological age was regressed on identified predictors, and the residual values (distance from the prediction line) were utilized to estimate biological age (Duggirala et al., 1992). The predictors represent various organs or physiological systems and are often associated with chronic diseases. Biological aging was assessed through a set of complex phenotypes: (1) neuromuscular—hand steadiness, grip strength, hand-eye coordination, reaction time, motor time, and trunk flexibility; (2) pulmonary function—FEV1 (one-second forced expiration volume) and FVC (forced vital capacity); (3) anthropometrics—BMI and four skinfolds (biceps, suprailiocristal, subscapular, and triceps skinfolds); (4) biochemical markers—panel of lipid profiles, liver enzymes, electrolytes, blood urea nitrogen (BUN), creatinine, blood proteins, and thyroxin (T4); and (5) physiology—systolic and diastolic blood pressures and heart rate. The 10 significant predictors of chronological age include biochemical measures, BUN, glucose, total cholesterol, albumin, and *ln* potassium. Functional measures that are statistically significant +/– 0.05 include systolic blood pressure, FEV1, reaction time, trunk flexibility, and grip strength. The genetic variance (heritability) based on residuals (the difference between chronological and predicted ages), biological age, is 0.469 +/– 0.180 with a p-value of 0.0053 (Duggirala et al., 2002). Roughly, one-half variation associated with biological aging in Goessel Mennonites is controlled by genetics and one-half by environmental factors (Duggirala et al., 2002). However, our estimates of heritability for biological age may be slightly inflated due to the influences of shared environment.

Multivariate analyses identified predictors for an individual on chronological age and use of residuals as the estimate of biological age. Predictors represent various organ or physiological systems, often associated with chronic diseases (Uttley, 1991; Uttley and Crawford, 1994). Creatinine and BUN are associated with kidney function and are major predictors of mortality. Glucose (diabetes) plays a major role in predicting survivorship. Study age for both males and females provides the best predictor for being alive or dead (Uttley and Crawford, 1994). Biological aging has a strong genetic determinant in Mennonite populations (Uttley, 1991; Duggirala et al., 2012).

2004 Field Follow-up

In 2004, funds became available through the Kansas Attorney General's Office to conduct follow-up research dealing with nutrition and health in populations of Kansas. These funds were collected by the State of Kansas from commercial nutritional companies who misrepresented their products. Doctoral candidate M. J. Mosher and I applied for a grant from the Attorney General's Office Settlement Fund on the Kansas Mennonites and received generous financial support ($123,000). We were able to organize field teams consisting of several graduate and undergraduate students from KU (M. J. Mosher, Mark Zlojutro [Kos], Rohina Rubicz, Kris Young, Phil Melton, and Ellen Quillen) plus a postdoctoral fellow from Argentina, Dario Demarchi. Ellen Quillen (an undergraduate student at KU) served as coordinator and welcomed each participant at the door, obtained signed informed consent, and guided the participant to the appropriate station. We hired phlebotomists from the local hospital to draw blood samples.

The funds from the Attorney General's Settlement grant permitted the expansion of the samples from the original three groups (Goessel, Henderson, and Meridian) to six populations. Figure 6.2 locates the six Mennonite populations on a map of Kansas and Nebraska. Based on origins and history, Goessel and Henderson (General Conference Mennonites) should be most similar genetically because they shared common ancestry until the 1870s when a community from Crimea, Russia, emigrated to the United States and underwent fission. Lone Tree and Garden View were offshoots of Halderman Meridian Mennonites that share common ancestry but a complex, heterogeneous origin. Old Order (Old Colony) Mennonites were transplants from Cuauhtemoc, Mexico, to western Kansas and had a unique history and ancestry. These Old Colony Mennonites originally came from Chortitza, Russia, and then settled in Manitoba, Canada, and eventually 7,000 emigrated to Mexico in 1926. A small group of Mennonites from Cuauhtemoc, Mexico, moved to western Kansas to escape the drug wars and severe violence of northern Mexico.

The Mennonite research program provided excellent training to both graduate and undergraduate students and postdoctoral fellows of the Anthropology Department at KU. Five graduate students wrote dissertations on topics ranging from population dynamics (inbreeding—Laurine Rogers, 1984), social anthropology (Jeff Longhofer, 1986), genetics of lipids (Ravi Duggirala, 1995), biological aging (Meredith Uttley, 1991), and genetics of blood pressure in Mennonites (Sobha Puppala, 2000). In addition, the reconstruction of large Mennonite families permitted the analysis for preparation of M.A. theses on the genetics of thyroxine (Lisa Martin, 1993), serum calcium, and acid phosphate levels of the blood (Alexa Pfeiffer). Postdoctoral fellows in the Laboratory of Biological Anthropology employed anthropometry to examine Mennonite body morphology (Paul Lin), the genetics of neuromuscular and pulmonary function (Eric Devor), nutrition (M. J. Mosher), and the genetics of apolipoproteins (initiated by Dario Demarchi).

DNA Results—MtDNA

Mitochondrial DNA and Y-chromosome haplotypes (groups of alleles that are uniparental—i.e., inherited from a single parent) were characterized in all Mennonite participants. A research assistant (Melton, 2010) first characterized the mitochondrial DNA haplotypes and haplogroups (groups of similar haplotypes that share common ancestry with a single nucleotide mutation) in 118 Mennonite participants. They exhibited eight common Western European haplogroups: H, HVO, I, J, K, T, U, and X (Melton, 2012). H was the most frequent mtDNA haplogroup in all six of the Mennonite communities—ranging from 35% among the Halderman community of Lone Tree to 75% in a smaller sample of Old Colony. Multidimensional scaling (MDS) and principal component analyses (PCA) based on mtDNA revealed as predicted a close genetic relationship between Goessel and Henderson, thus reflecting the fission of the Crimean Mennonites into these two communities on arrival to the United States. The Halderman Mennonite communities, Garden View, Meridian, and Lone Tree, cluster together and reflect the fission from ancestral population, Meridian. The molecular genetic data support the historical relationship between the Mennonite communities and are indicative of patterns of fission-fusion along familial lines. All of the analyses indicate the genetic uniqueness of the Old Colony Mennonites based on history and small sample sizes.

Y-Chromosome Markers

Several common European Y-haplogroups were identified among the six Mennonite communities, including haplogroups R1b (56.3%), R1a and I2 (9.6%), E1b (6.4%), I1 and Q (5.3%), G2a (3.2%), J2 (2.1%), L, and an unidentified haplotype (1.1%). The distribution of these haplogroups varied among the congregations (see Figure 6.6). Haplogroups R1b (50%–63.2%) and R1a (4%–26.3%) were the only haplogroups found in all five congregations (Beaty 2016). The Alexanderwohl congregations of Goessel and Henderson both exhibited haplotypes belonging to haplogroups G2a and J2a. Goessel displayed the most haplogroup diversity with six haplogroups—R1b (53.8%), I2a (15.4%), J2a, G2a, R1a, and Q (7.7%)—represented in 13 individuals and the highest mean number of pairwise differences between haplotypes (11.48). Meridian exhibited the second highest haplogroup diversity with

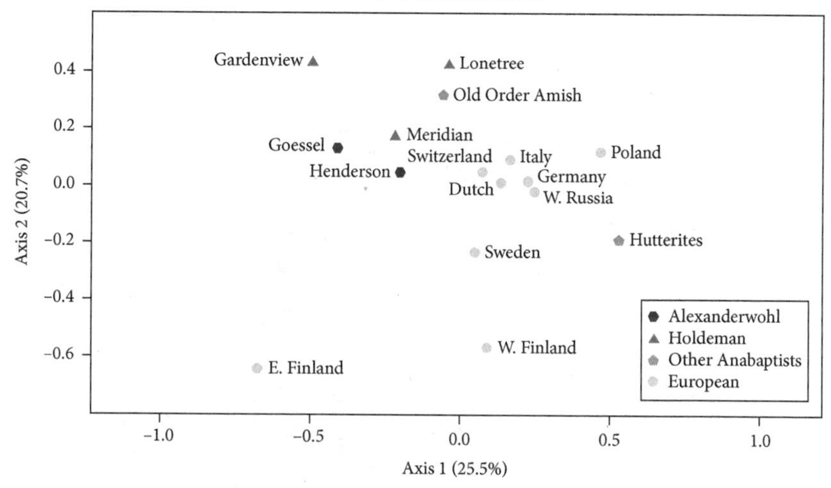

Figure 6.6 Y-chromosome haplotype frequencies in five Mennonite communities.

Source: Adapted with permission from Beaty, K. G. et al., "Paternal Genetic Structure in Contemporary Mennonite Communities from the American Midwest," *Human Biology* 88(2): 95–108. © 2016, The Authors.

at least 7 NRY haplogroups present (R1b, 56%; E1b1, 12%; I1a and Q, 4%; I2, L, R1a, and one unidentified haplotype, 4%) in 25 individuals and an average number of pairwise differences of 11.22. It was the only congregation to exhibit haplogroup L. Lone Tree exhibited the lowest haplogroup diversity, with 63.2% of the individuals belonging to haplogroup R1b, 26.3% belonging to haplogroup R1a, and 5.3% representing haplogroups E1b1a and I2a; whereas Garden View had four haplogroups represented (R1b, 50%; I2, 35.7%; R1b and Q, 7.1%) and exhibited the lowest average number of pairwise differences between haplotypes, 9.8476 (Melton, 2016).

Population Structure

AMOVA results for NRY STR and mtDNA variation are summarized in Table 6.2. As with mtDNA, the magnitude of variation seen among groups is lower than seen within all populations, while the amount of variation within populations is high. The amount of variation seen within populations is higher for mtDNA versus NRY STR data. The amount of NRY STR

Table 6.2 Results from Stepwise Discriminant Function of Blood Chemicals and Neuromuscular Performance

Group	Classified Correctly (%)	Alive	Deceased	Group Centroid	Variable	Factor Loadings*
Women					Study age	−0.73278
Alive	79.4	370	96	0.12665	BUN	−0.66957
Deceased	80.3	24	57	−1.59657	BUN/creatinine ratio	−0.50659
					Chloride	0.35658
					LDL	−0.31418
					Creatinine	−0.30010
					Albumin	0.27958
Men					Study age	0.78119
Alive	79.9	321	81	−0.20305	BUN	0.47637
Deceased	78.5	14	51	1.51880	Albumin	−0.43012
					Glucose	0.35172
					BUN/creatinine ratio	0.34564
					Albumin/globulin	−0.27615
					Total protein	−0.18944

BUN, blood urea nitrogen.

*Correlation between the variable and the discriminant function standardized by the pooled within-group variances.

Source: Reproduced with permission from Uttley, M., *Relationship of Measures of Biological Age to Survivorship among Mennonites*. Ph.D. diss., University of Kansas, 1991.

variation explained among communities within each grouping is lower (6.12% vs. 7.57%) when Lone Tree is treated as a separate group from the other Holderman communities. When Mantel tests are applied to these data, geographic proximity, geography, and NRY variation are negatively correlated with mtDNA variation (Table 6.2). This is particularly true of NRY STR and mtDNA distances ($r = -0.5531$); however, neither of these results gave significant p-values. There is a slight correlation ($r = 0.1283$) of geography by NRY STR distances, but these results are also nonsignificant.

Intrapopulation Analyses

Principal components analysis (PCA) examined the relationship of the Mennonite congregations with other Anabaptists groups based on a reduced number of loci. The PCA of Mennonite communities plots 77.5% of the NRY variation on the first two axes and shows that these Mennonite congregations are more similar to each other than they are to either the Old Order Amish or Hutterite Anabaptist populations (see Figure 6.6). However, this PCA also indicates that Mennonite communities do not cluster with their founding congregations of Alexanderwohl and Holdeman, while the community of Garden View is the farthest from the Mennonite communities in the plot. There is no sharing of paternal haplotypes between the Mennonite communities based on an expanded set of 15 NRY STRs, but with a reduced loci set of 6 STRs, sharing of haplotypes does occur. All of the shared paternal haplotypes between the communities belong to the most common and widespread western European NRY haplogroup, R1b. The sharing of paternal haplotypes also includes 4 R1b haplotypes shared with the Old Order Amish, with at least one of these haplotypes found in each of the Mennonite communities. However, no Mennonite communities shared haplotypes with Hutterite populations included in this analysis, and this group is the furthest outlier within this plot (Beaty et al, 2016).

Population genetic studies examining religious isolates are particularly informative for the study of rare genetic disorders, due to their unique population structure (Pollin et al., 2007; Pichler et al., 2010; Melton et al., 2006). However, to date few studies have examined molecular genetic data to interpret the diaspora of Anabaptist populations following the Reformation. The one focus for this current study was to determine the paternal genetic relationship between five Mennonite communities using NRY STR data and to determine if these data support the history of fission-fusion that has been demonstrated for these groups. Despite a shared history originating after the Reformation, there is no sharing of paternal haplotypes between Mennonite communities. Figure 6.6 shows the close genetic affinities between Henderson and Goessel based on Y-chromosome haplotypes. West European haplotype R1b is by far the most common in all of the Mennonite communities. Meridian, Garden view, and Lone Tree, all related Halderman populations, show greater variability that reflects population fission from the founding Meridian population. Using PCoA, Figure 6.6 compares Mennonite populations to other European groups and shows their close genetic affinities to German, Dutch, and Swiss populations.

In addition to the uniparental markers being utilized to characterize the genetic structure of Mennonite populations, a postdoctoral fellow from Argentina (Dr. Dario Demarchi et al., 2005) identified Apoprotein genetic variation for APOE and APOB. However, these genetic markers are less polymorphic and the allelic frequencies appear to be less useful in the reconstruction of the evolutionary history of the Mennonite congregations (Demarchi et al., 2005).

Mennonite Anthropometrics and Neuromuscular Function

Anthropometric variation results from the interaction of complex genetic factors and environmental interaction. This research program permitted a comparison of two groups that resulted from the fission of a single congregation almost 100 years ago (1874–1979). These two related groups (Alexanderwohl and Bethesda) were compared to Meridian Mennonites who arose from the combination of several Anabaptist groups.

A total of 1,197 Mennonites (567 males and 630 females) from Alexanderwohl and Meridian, Kansas, and Henderson, Nebraska, volunteered to be measured anthropometrically. Thirty-five different anthropometric traits were utilized to characterize the morphological variation observed in Mennonite populations (Devor et al., 1986a). All of the same anthropometric measurements were taken by each of the three researchers (Paul Lin, Laurine Oberdieck Rogers, and Joanne McGreevy). In this way, interobserver error was minimized, with each researcher measuring all subjects for specific traits.

The effects of age and sex on neuromuscular performance were examined in a cross-sectional sample of 559 members of the (Alexanderwohl) Goessel community. Age and sex effects on neuromuscular traits were assessed using a stepwise polynomial regression method. Of the six neuromuscular traits (DHS, HST, HEY, SRT, MT, and TKF) only hand steadiness (HST) failed to show a significant sex difference. Trunk flexibility (TKF) did not show a significant nonlinear trend with age (Figure 6.5). All of the other neuromuscular traits display a decline in performance after age 45 of up to 60% (Devor et al., 1985).

Conclusion

Biological aging indices provide a more accurate predictor of survivorship than chronological age among the Mennonites. Because of the longitudinal nature of this research, the relationship between aging and mortality was tested. The stepwise discriminant function computes one or more vectors that maximally separate two or more groups or subjects. Discriminant analysis computes an individual function score that is used to predict group membership of individuals into alive versus deceased status. Males who aged faster than the predicted rate (by stepwise multiple regression analysis) died sooner than those who were "slower agers" (Crawford, 2005).

Despite the relatively recent subpopulation fission, populations of Mennonites have differentiated genetically. The observed genetic differentiation was caused by the genetic makeup of the founding populations, genetic drift, and patterns of gene flow.

7

Genetic Structure and Origins of Siberian and Alaskan Arctic Populations

Introduction

In 1976, I received a grant from National Science Foundation (NSF; OPP-0327676) for initiating a research program on the peopling of the Americas based on a comparison of genetic markers of the blood from Siberian and Alaskan Native populations. This research program was designed to compare the genetics of Yupik-speaking Alaskan populations with two groups from Kamchatka, Siberia. Unfortunately, at that time, the Soviet government did not approve field research by American scientists on indigenous populations of Chukotka. Since the original experimental design was not possible, my alternative was to compare two Yupik-speaking Eskimo populations of St. Lawrence Island (Savoonga and Gambell) with two Inupik-speaking populations of Wales and King Island, Alaska (Crawford, 2007e). The gene frequencies from these four populations were compared with those of Siberian groups located on the coast of Chukotka (allelic frequencies compiled from Russian literature). Norton Sound Health Corporation of Nome, Alaska, and tribal organizations in each community approved the Alaskan portion of the study. Because of a particularly "frigid period" in the Cold War, we could not get approval from the Soviet government to conduct any comparative research on the indigenous populations of Siberia. With NSF approval, the proposed research program excluded field investigations in Siberia. Instead, I focused on a genetic comparison of Yupik versus Inupik-speaking Eskimo populations and compared them to published Siberian and Greenland gene frequencies. Such a comparison limited the number of genetic markers available for population comparisons because at that time Soviet genetic laboratories phenotyped fewer genetic markers than US facilities.

A total of 250 Yupik-speaking Inuit volunteers from St. Lawrence Island (170 from Savoonga and 80 from Gambell) participated in this study. Before

In Search of Human Evolution. Michael H. Crawford, Oxford University Press.
DOI: 10.1093/9780197679432.003.0007

1878, St. Lawrence Island population consisted of approximately 4,000 inhabitants distributed among 35 settlements located around the periphery of the island (Crawford et al., 1997). By 1904, European whalers and traders moved to the island, bringing disease and famine, while reducing the population from an estimated 4,000 to 250 persons. In 1900, the US government introduced reindeer domestication and herding to the populations of St. Lawrence Island. This resulted in the formation of two separate villages, Gambell and Savoonga (Byard and Crawford, 1991). At present, approximately 400 persons continue to reside in each community.

The village of Wales, located at the narrowest point of separation of Asia and Alaska, contained the largest known Eskimo village in historical Alaska (see Figure 7.1). This village had an estimated size of 400 Inupik speakers, but the total population size was reduced to 117 persons by 1975. This reduction in size was due to the introduction of diseases, such as the influenza epidemic of 1918 plus some outmigration.

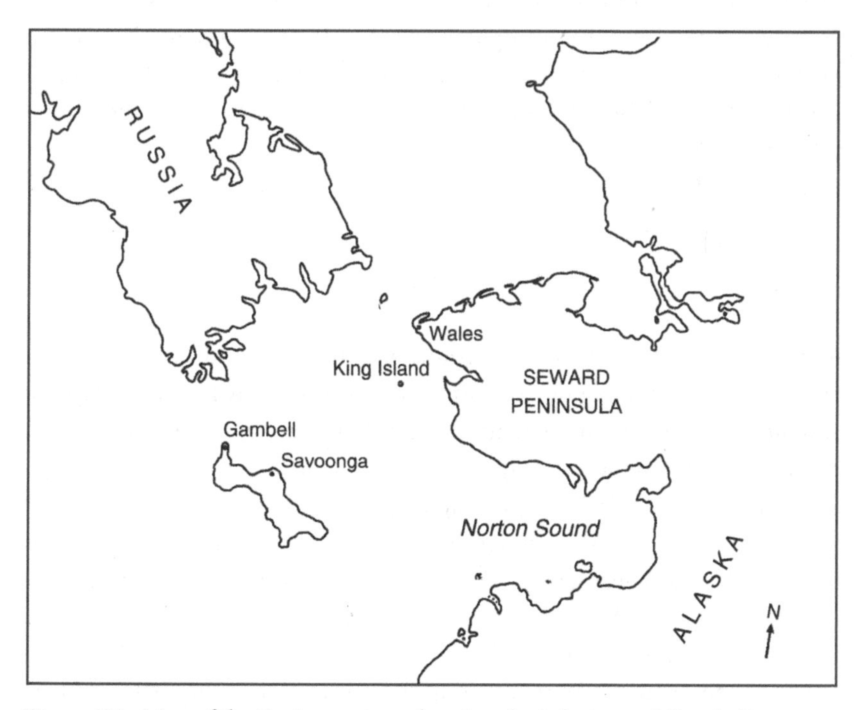

Figure 7.1 Map of the Bering region, showing the locations of Gambell, Savoonga, King Island, and Wales, Alaska.

Source: Reproduced with permission from Crawford, M. H., *The Origins of Native Americans*. p. 44, fig. 14. © 1998, Cambridge University Press.

King Island is a small island located due south of Wales, Alaska, and west of the Seward Peninsula (see Figure 7.1). In 1791, its population size, based on historical records, was estimated to be 170 persons. However, this community relocated in its entirety to Nome, Alaska, with 215 King Islanders currently residing in Nome.

Participants

Field Research

During the summer of 1978, a field research team (consisting of Pamela Byard, Paul Lin, Michael Crawford, and dentist Terry Rigdon) from the University of Kansas spent one month on St. Lawrence Island, Wales, and Nome, Alaska, collecting demographic, morphologic, and genetic data from the four indigenous communities, containing 1,120 residents. A total of 370 participants (one-third of the entire population) agreed to participate in the study and to submit to venipuncture. The blood specimens, drawn into vacutainers containing ACD preservative, were packed in ice and shipped to the War Memorial Blood Bank of Minneapolis for analysis of standard genetic markers.

Genetic Analyses

Allelic frequencies of 13 blood groups and 10 red blood cell and serum protein loci for four populations were compiled in Crawford et al. (1981). Seventeen of these loci are polymorphic, and three are monomorphic. A plot of a three-dimensional least-squares "genetic map" clearly reflects the history of the four populations with Yupik-speaking Gambell and Savoonga of St. Lawrence Island clustering closely together, while Inupik-speaking Wales and King Island reflect genetic differentiation (Byard and Crawford, 1991; Crawford et al., 1981).

A pseudo-three-dimensional plot of 19 circumpolar groups based on standard genetic markers reveals that there is a relationship between language and genetic affinities. Wales and King Island form a tight cluster together with other Inupik-speaking populations—Wainwright, Barrow, Anaktuvuk Pass, and Greenland (see Figure 7.2). The second, more diffuse

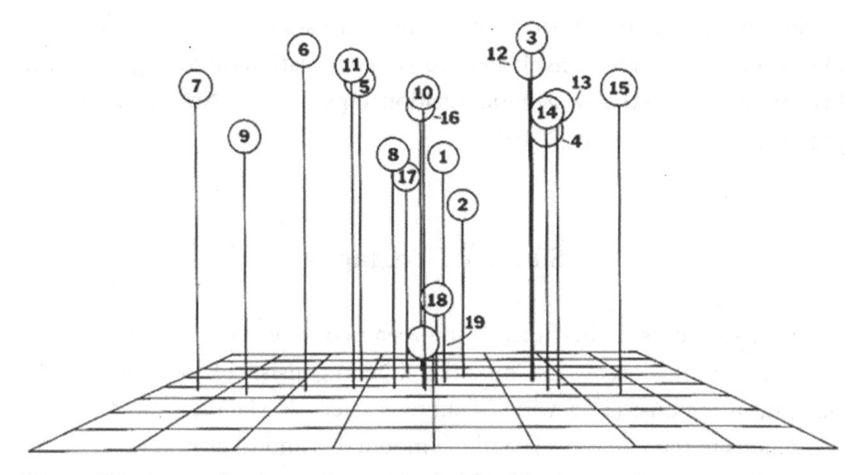

Figure 7.2 A pseudo-three-dimensional plot of the genetic structure of 19 circumpolar populations.

Source: Adapted with permission from Crawford, M. H., Mielke, J. H., Devor, E. J., Dykes, D. D., and Polesky, H. F., "Population Structure of Alaskan and Siberian Indigenous Communities," *American Journal of Physical Anthropology*, 55: 168, fig. 1. © 1981, Wiley.

cluster includes Siberian and Yupik-speaking communities. Samoyed groups located on the Tamyr Peninsula of Siberia are clearly separated along the second axis from all of the other populations. Aleutian Bering and Mednii Islands cluster with the Siberian populations, supporting their historical origins. There is considerable agreement in the genetic affinities between Siberian and Alaskan populations as revealed by blood groups and proteins.

Recent European admixture and resettlement programs have added some complications to the interpretation of the genetic structure of the four Alaskan Eskimo populations. Admixture was computed using the incidence of immunoglobulin haplotype $Gm^{f;b}$, a marker haplotype found primarily in Europeans. This genetic marker was entirely absent in the King Island population but revealed that Gambell on St. Lawrence Island had 8.8% European ancestry, while Savoonga displayed 4% European ancestry (Byard et al., 1983). Wales, Alaska, exhibited an incidence of $Gm^{f;b}$ of 0.23 and an estimated European contribution of 7.6% (Crawford, 2007, 2010a). Eskimo populations from St. Lawrence Island (Gambell and Savoonga) and Pribilof Island Aleuts (St. Paul and St. George, Alaska) were compared anthropometrically, using data collected by William Laughlin from 1979 to 1981 (Justice et al., 2010). Anthropometric measurement was used to reconstruct population history of Aleuts and Eskimos of the Bering Sea. Our results show a relationship between the populations linking geography, history, and underlying genetic relationships among Native American populations (Justice et al., 2010).

Siberian Geography

Siberia encompasses one-third of the Asian continent with 11 time zones stretching from Moscow to Chukotka. Siberia consists of two unequal geographical regions: (1) a wide, swampy, eastern lowlands, from the Ural Mountains to the Lena River; and (2) mountains and river valleys from the Lena River to the Pacific Ocean. The ecology of Siberia consists of the taiga, tundra, and steppe grasslands of the south. Tundra is a flat moss- and lichen-covered expanse characterized by poor drainage, a short growing season, and the general presence of permafrost. The Taiga is an expanse made up of mixed deciduous and coniferous forest. The steppes of Southern Siberia consist primarily of grasslands.

The widely distributed populations of Siberia cluster into three different linguistic groups: (1) Altaic speakers; (2) Uralic speakers; and (3) Paleo-Asiatic speakers. The Altaic speakers include Tungusic groups, such as the Evens, Evenki, and Udegeys. The Tungusic populations such as the Evenki reindeer herders number 29,900, Eveni speakers 17,000, while the Udegeys are the smallest, most isolated group consisting of fewer than 1,900 persons and located in far eastern Siberia, near the Sea of Okhost. Numerically, the largest populations of indigenous Siberians consist of Uralic Samoyedic speakers: Entsi (N = 190), Nentsi (N = 34,200), Ngansani (N = 1,300), and Sel'kupi (N = 3,600). Included in the Uralic speakers are Ugric groups, consisting of Khants (N = 22,300) and Mantsi (N = 8,300). Paleo-Asiatic speakers include the populations of the Kamchatkan Peninsula: Chukchi (N = 15,100), Itel'meni (N = 2,400), Koryaks (N = 8,900), and Eskimo-Aleut (Yupik Eskimos, N = 1,700, Aleuts, N = 650). Of special anthropological interest are the linguistic isolates: Kets (N = 1,080), Nivkhs (N = 4,600), and Yukaghirs (N = 1,000).

Based on linguistic evidence, several studies have proposed that the Kets are related to Na-Dene speakers of Canada and must have had recent common ancestry (Ruhlen, 1991, 1998; Vayda, 2010). However, genetic and DNA data do not support this hypothesis (Rubicz et al., 2002).

Russian Contact

Russian contact with indigenous populations of Siberia began in 1580 when Yarmak (an outlawed Cossack chieftain) crossed the Urals with fewer than 2,000 men and conquered the lands of Tartar prince Kutshum Khan. The capital of these conquered Tatar lands was named "Sibir." Russians advanced river valley to river valley founding towns as they moved in an eastwardly direction. Tobolsk was founded in 1587, Yakutsk in 1632, and Irkutsk in 1651. The first Russian government expedition to Kamchatka was organized in 1733.

Siberia, At Last!

In 1991, we finally received permission through the Institute of Cytology and Genetics, Russian Academy of Sciences at Novosibirsk, to begin a research program with indigenous populations of Siberia. In 1958, Demitry Konstantinovich Belyaev transplanted this academic institute from Moscow to Academgorodok in Siberia. Because of his acceptance of the concept of Mendelian genetics, Belyaev was dismissed from the Department of Fur Animal Breeding. The relocation of the Institute from Moscow to Siberia permitted Belyaev to escape the political quagmire and intrigues of Moscow. At that time, Denisovich Lysenko, a Soviet agronomist, had rejected Mendelian genetics in favor of Lamarckism, which better fit Soviet ideology. Stalin had appointed Lysenko in 1940 to the directorship of the Institute of Genetics in Moscow.

Academician Demitry Belyaev, together with Dr. Ludmila Trut, initiated the experimental behavioral genetic study of silver foxes. In 1952, they began domesticating silver foxes at the Institute of Genetics, Academgorodok, in Novosibirsk. To date, there are more than 60 generations of silver foxes raised at the Institute. I was fortunate to receive the grand tour of the facilities and be given a historical explanation of their research by Professor Demitry Belyaev.

In the midst of our field investigation on the genetics of a lymphoma epidemic among the baboons of Sukhumi, Abkhasia, Dennis O'Rourke and I were informed by Dr. Rem Sukernik (director of the Genetics Laboratory, Institute of Cytology) that permission was granted for field research on the Evenki reindeer herders of Central Siberia (see Figure 7.3). Politically, this was a tense time in Sukhumi prior to the war between Abkhasia and Georgia. We were invited

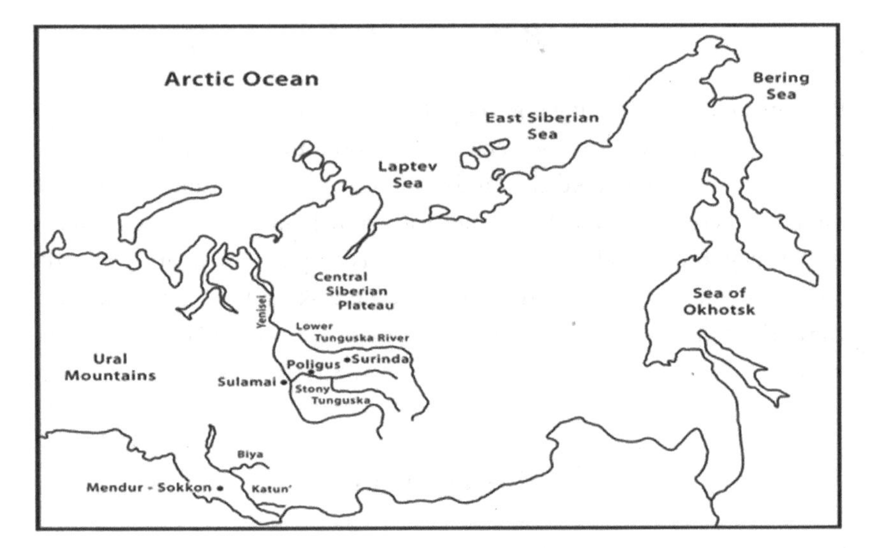

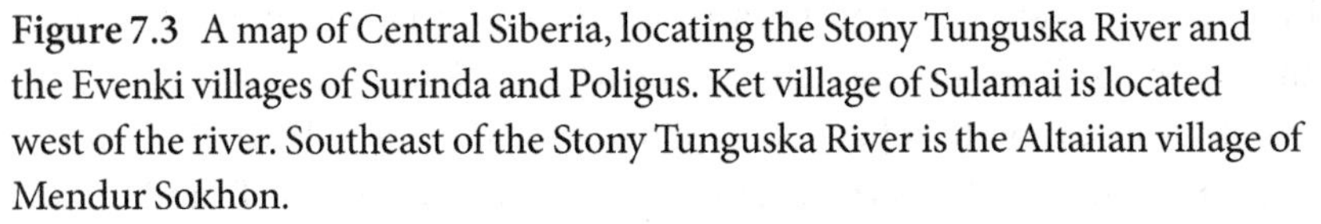

Figure 7.3 A map of Central Siberia, locating the Stony Tunguska River and the Evenki villages of Surinda and Poligus. Ket village of Sulamai is located west of the river. Southeast of the Stony Tunguska River is the Altaiian village of Mendur Sokhon.

Source: Adapted with permission from Crawford, M. H., Mielke, J. H., Devor, E. J., Dykes, D. D., and Polesky, H. F., "Population Structure of Alaskan and Siberian Indigenous Communities," *American Journal of Physical Anthropology*, 55: 180, fig. 6. © 1981, Wiley.

by the Academy of Sciences to join Rem Sukernik in Academgorodok, and he announced that we could fly north from Sukhumi to Moscow and Krasnoyarsk and then proceed to Evenkiya for a pilot study. We arrived by Aeroflot jet plane to Karsnoyarsk, which was officially a restricted zone because of its role in the production of nuclear weapons. The officials at the airport who did not expect us "freaked out" when we arrived, but Sukernik was able to convince them that we had governmental permission to travel to Evenkiya. From Krasnoyarsk, we flew on a small propeller-driven Aeroflot plane to the town of Baykit, a small town of 3,500 persons located along the Stony Tunguska River. For this occasion, the pilots prepared a "picnic lunch" consisting of beluga caviar, smoked fish, and champagne. During the flight, they organized a Siberian "picnic" on the floor of the airplane.

The pilots were thrilled to be able to transport an American research team. Since I was the only team member besides Russian native Rem Sukernik who spoke Russian, I was seated in the front cabin next to the pilot. Midway through the flight, the pilot inquired if I would like to take over the controls of the plane. I responded by explaining that I had never flown a plane. The pilot reviewed the simplicity of the controls and handed them over to me. In the back of the airplane, the research team became nervous at the thought of a novice flying the plane. As we approached the runway, the appropriate pilot took control of the airplane and expertly landed it. This plane was greeted by a "canine delegation," which chased it to a stop.

You Gotta Have Heart!

Surinda

Since there were no roads leading into this region of Evenkiya, our only option was to travel by plane or helicopter, rented from Aeroflot. On the expedition in 1992, we planned to visit two of the Evenki reindeer herding brigades: Surinda and Poligus. We boarded the helicopter in the town of Baykit and flew across the taiga to a clearing on the edge of the Surinda brigade. We were greeted by the local headman who guided us with all of our gear up a steep hill to a settlement of chums (similar in appearance to Midwestern Native American tepees) that housed the families of the herders. Since we were the first westerners to visit the brigade, the chief had a reindeer slaughtered in our honor. As the most senior visitor and director of our team, a still-beating heart of the reindeer was first offered to me. I thanked the headman profusely but explained that because of my age, I was dyspeptic and could not digest such a special "treat"; however, the young guys of my team loved raw heart! Unable to object in Russian, Tony Comuzzie and Bill Leonard were good sports and accepted this cardiac delicacy. As they were sitting on a log, chewing on bits of the heart, I asked them how it tasted. Bill, with blood dripping from his mouth, cheerfully replied: "It's a little tough but has the texture of bubble gum." However, Tony quietly munched on his cardiac treat. That evening we were invited to an outdoor dinner arranged by the headman (see Figure 7.4). As our group settled down around a blanket adorned with Evenki delicacies, clouds of Anopheles mosquitos announced that we were on their menu. American veins are a rare treat in Evenkiya, and the mosquitos attempted to feed on our blood. Fortunately, I wore a hat with a mosquito net and managed to raise it periodically for an occasional insertion of food. I spotted one of my favorite Siberian foods, deep-fried sour dough pastry normally containing ground beef inside. My mother (born in Siberia) used to prepare these "peroshki" stuffed with ground beef, onions, and eggs. I bit into one of the Evenki peroshki and discovered a crunchy,

Figure 7.4 Dinner in Surinda hosted by the chief of the Evenki community. Included in this photograph are members of the research team: A. Comuzzie, R. Sukernik, M. H. Crawford, and community head.

unique, and somewhat unpleasant taste that definitely did not have the taste and texture of ground beef. I naively inquired: "What's in the peroshki?" The headman explained that the Evenki use chopped reindeer lung tissue for the peroshki. That was the last peroshok that I ate while in Siberia.

Throughout this outdoor dinner, our team physician Rem Sukernik was busily swatting mosquitos off his forehead. Blood was oozing from the bites, so I offered the spare mosquito net to him. He refused the offer, emphatically stating: "You cannot let it bother you! Ignore the mosquitos." Even though it was more difficult to eat with a mosquito net draped across my face, I refused to follow his instructions and kept the mosquitos away using my methods (see Figure 7.4).

Poligus

After sampling the population of the Surinda brigade, the research team boarded an Aeroflot helicopter and flew to the adjoining brigade, Poligus.

As we hovered over a clearing, I noticed human bodies scattered all over the ground. As we landed, these "bodies" slowly stirred, arose, and shuffled toward the helicopter. We learned that Poligus had received a huge shipment of vodka, from Japanese traders, in exchange for reindeer antlers. These antlers were highly prized in Asia as aphrodisiacs and often traded for alcohol and TV sets—even though there was no available electricity in this area. The males from the community had been heavily drinking vodka for several days until they passed out, and many of the men collapsed in a drunken stupor in the clearing. Our arrival by helicopter revived the drinkers, and after we deplaned, they demanded that we fly them to the adjoining brigade who had recently received a shipment of vodka. The revelers forced their way into the helicopter, prompting the Russian crew to push them out of the rear exit. After several confrontations, one of the Evenki males took out his knife and slashed a helicopter tire. Given the dangers that one flat tire poses—it makes the helicopter list to one side, thus making takeoff and future landings precarious—we would be stuck in the middle of the taiga, hundreds of miles from the nearest town. We immediately loaded ourselves back into the helicopter and took off. Since one of the tires was flat, we hovered a few feet over the clearances in the forest. One of the pilots jumped out of the helicopter, dragged a log from the forest, and placed it under the flat tire. In this way, through the bravery and skill of the Russian pilots, we managed to continue visiting the various brigades.

Before the Soviet takeover of the Siberian taiga, each extended Evenki family controlled and grazed its own reindeer herd. With the Russian communist reorganization, extended families were forced to join these so-called brigades. In the early spring, representatives of each brigade searched for new grazing for their reindeer herds, and when appropriate grounds were located, the brigade members would drive the herds from their winter locations to the rich grasslands of the taiga (see Figure 7.5). Each of the brigades herded approximately 2,000 reindeer. Normally, during the winter months the Evenki children attend boarding schools in towns such as Baiykit. Thus, the Evenki speak their native Tungusic language but are literate in the Russian language. During the summers, reindeer herding families are reunited with their children in the brigades located in the taiga. Figure 7.6 shows the similarities in the construction between Evenki chums and tepees of Plains Indians in the United States.

Figure 7.5 Evenki reindeer herder saddled atop a reindeer.

Kets of Sulamai

The Kets are numerically one of the smallest indigenous populations of Siberia, with fewer than 1,000 persons living along the Yenissey River. Their primary form of subsistence is fishing, supplemented by hunting. The last of the surviving Yenesain speakers were located in a small village next to the Yenesai River. Bill Leonard, Rem Sukernik, and I rented a small boat and crossed the wide Yenesai River to the village of Sulamai. This village

Figure 7.6 Evenki housing.

consisted of a small number of wooden huts, and we set up to collect blood from as many volunteers as possible. A total of 25 Kets agreed to participate and donated blood for this study. Studies on the classification of languages of Siberia by F. Merritt Ruhlen suggested that NaDene languages cluster with Yenisean languages. This conclusion was further supported by Edward Vayda's research based on Ket vocabulary. He proposed the existence of a close linguistic affinity between the Ket speakers of Siberia and Na-Dene speakers of Alaska despite a geographic separation of a thousand miles (Vayda, 2010). However, a genetic comparison between Ket and Na-Dene populations did not reveal such a close affinity. Instead, the Kets clustered with central Siberian populations, while the Algonkian groups clustered with Canadian indigenous groups (Rubicz et al., 2002).

Altai—Kizhi of Gorno Altai

In the summer of 1983, a research team from the Siberian branch of the Russian Academy of Sciences drove from Novosibirsk to Mindar-Sokhon, to sample the Turkic-speaking, cattle-herding, Kizhi population of the mountain Altai. The research team included Drs. Tatiana Karafet, Olga Posukh, Michael Crawford, and Ludmila Osipova. Our collaboration with Rem Sukernik had been terminated because (1) he was illegally peddling the Siberian blood samples that we collected to various US laboratories; and (2) he insisted that my NSF grant should continue to support travel and subsistence in the United States for Elena Starikovskaya, even though she had completed training in mtDNA phenotyping in Doug Wallace's laboratory. Earlier, I had arranged for Sukernik to receive similar training in molecular genetics in Mark Stoneking's laboratory at Penn State University. However, shortly after Sukernik's arrival at Stoneking laboratory, I received a telephone call from Stoneking requesting that I take Sukernik off his hands. Apparently, Sukernik refused to master the basic laboratory procedures required for a collaborative project. Fortunately, Dr. Wallace was willing to take him into his laboratory in exchange for access to the blood specimens. That is how Sukernik ended up in Atlanta at Emory University (Torroni et al., 1993).

Results

A comparison of mtDNA variation of the Evenki and nine other Siberian indigenous populations with Native Americans revealed the presence of haplogroups A, C, and D, but the absence of haplogroup B (Toroni et al., 1993). Analyses included PCR amplification, restriction analysis, and control region sequencing (Comuzzie, 1993). Table 8.1 in Chapter 8 summarizes the distribution of mtDNA haplogroups in the Bering region of Siberia/Alaska. Haplogroup B is defined by a 9 bp deletion in region V of the mitochondrial DNA. The presence of haplogroups A, C, and D support the age-old belief that Native Americans came from Siberia (Crawford et al., 1981). Comuzie and Crawford (1990) revealed the relationship between biochemical heterozygosity and morphological variability in Siberian populations (Comuzzie and Crawford, 1990).

Five Native American haplotypes, originated in Asia, with A, C, and D haplotypes distributed throughout Siberia. However, Siberian populations exhibit additional haplotypes such as F, G, M, U, and Z that have not been identified among Native Americans (Starikovskaya et al., 1998). Although haplogroup B is rare among Siberian populations, it has been detected among the Altai, the Buryats, and the Touvans (Kolman et al., 1996). Similarly, haplogroup X is infrequent among Siberian populations but occurs in the Altai at low frequencies. The presence of all five of the mtDNA haplogroups in the Altai is suggestive of Native American origins in southern Siberia (Rubicz and Crawford, 2016).

Y-Chromosome Markers

Y-chromosome markers are less variable than mitochondrial haplogroups and sequences. The majority of Native American populations belong to haplogroup Q, identified by a single nucleotide polymorphism (SNP) M242 (Seielstad et al., 2003). In addition to the presence of this SNP throughout the Americas, it has been identified in Asian, European, and North African populations. The subhaplogroup Q3 has been observed throughout the Americas and among the Siberian Inuits (Karafet et al., 2008).

Figure 8.5 in Chapter 8 plots the relationship between mean per locus heterozygosity and the distance from the centroid of distribution. Harpending and Ward (1982) developed a model that aids in the assessment of the effects of different micro-evolutionary forces. Deviation from the expected heterozygosity is suggestive of gene flow. McComb et al. (1996) applied the Harpending-Ward model based on VNTR variation in Siberian indigenous populations.

Other Molecular Markers—ABO

In addition to the standard uniparental markers, researchers from the Laboratory of Biological Anthropology examined the region of the genome (ninth chromosome) that codes for the ABO locus. DNA samples used in these analyses were collected during fieldwork in Siberia by M. H. Crawford (director of the LBA) and Larissa Tarskaya, a Siberian Yakut postdoctoral fellow at the LBA. In 2014, molecular and statistical analyses of these Siberian DNA samples were completed by graduate student Jacob T. Boyd, in fulfillment of his M.A. thesis. He is currently a medical resident at the University of Colorado Medical School.

A geographic map representing Siberian and Native American populations and their ABO haplotype frequencies is presented in Figure 7.7. All populations shown exhibit three O haplotypes: O1, O1v, and O1v542. Haplotype O1v was the most common, and O1v542 occurs in frequencies between 1% and 4% in all populations.

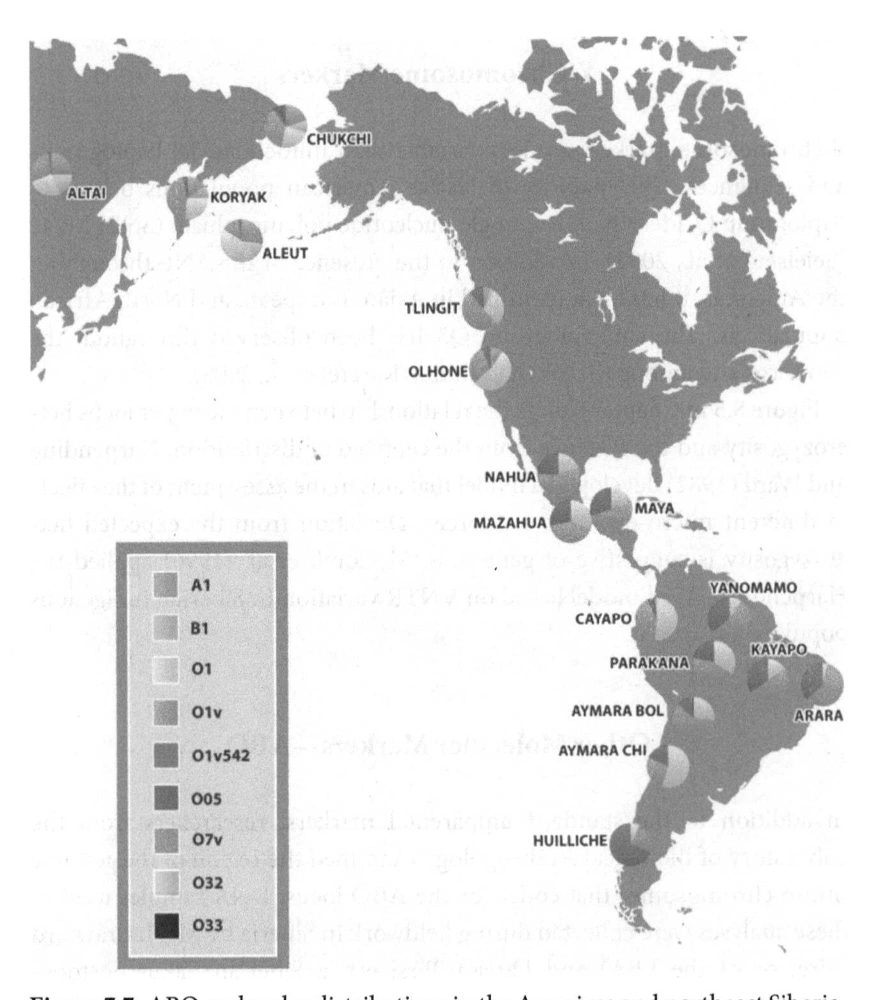

Figure 7.7 ABO molecular distributions in the Americas and northeast Siberia.
Source: Reproduced with permission from Boyd, J. T., *Peopling of the Americas: ABO Blood Group Haplotypes as an Indicator of Native American Origins and Migration from Siberia*. M.A. Thesis, University of Kansas, 2014.

The most frequent haplotype of the O blood group in the Americas is O1v (Estrada-Mena et al., 2010). Another haplotype differing from O1v by a single mutation, O1v542, has been discovered at frequencies between 4% and 60% in Native American populations but is absent in Asian populations. Researchers have utilized this marker as an ancestral informative marker (AIM) because of its broad distribution from Canadian Native American populations to Chilean populations of South America. Other AIMs have been observed in indigenous Native Americans, most notably the nine-base repeat allele (9RA), which along with mitochondrial DNA (mtDNA) and nonrecombining Y DNA (NRY-DNA), has led to the reconstruction of a likely Beringian route into the Americas from Siberia (Schroeder et al2009).

The Bering Incubation Model (BIM) hypothesizes that between 28,000 and 11,000 years ago an ancestral population moved across northeastern Siberia and then paused on the Bering land bridge (Meltzer, 2009). The combination of mutations and genetic drift resulted in unique marker development before continuing into the Americas. This expanding population may have stalled in Beringia because most of North America (from east to west) was covered by two ice sheets until approximately 16,000 BP (Lambeck et al., 2002).

Reich (2012) suggested a three-wave migration model from Siberia to the Americas. According to this three-wave model, a west coast migration across the Americas occurred first because that is where the ice sheets initially receded. The second wave moved inland between the two ice sheets. The third wave populated the northern latitudes, resulting in the establishment of Aleut, Inuit, and Eskimo populations.

These molecular analyses of the ABO blood group system in four different Siberian populations (Aleut, Chukchi, Koryaks, and Altai) were completed by Jacob Boyd (2014). The four populations from Siberia were associated with the original peopling of the Americas. The Aleuts from Siberia constituted the last migration into the Americas following the Beringian incubation (Crawford et al., 2010). The Chukchi and the Koryak, currently located on the Siberian side of the Bering Sea, are the two populations most consistently linked to Native Americans through genetic analyses. The Altai are a native central Siberian population who exhibit all five of the mtDNA haplogroups present in Native American populations. The four Siberian groups display all or many of the mtDNA (A, B, C, D) and NRY-DNA (C, P, Q, R) haplogroups found in the Americas along with the autosomal STR 9RA marker in Chukchi, Koryak, and Aleut populations (Crawford, 2007). This application of O1v542 as an AIM may also reveal if the frequency of the O blood group is the result of natural selection or genetic drift.

The presence of O1v542 in the Chukchi and Koryak populations is most likely the result of an expansion from an ancestral population and not due to back migration, thus supporting the proposed three-wave model of Tamm et al. (2007). The primary argument based on DNA data by Rubicz et al. (2010) against Kamchatkans originating from the same ancestral population as Native Americans is genetic discontinuity with the Aleut population, thus supporting back migration as an explanation for the observed frequency of 9RA. The Aleuts have a more recent migratory history (circa 9000 BP) compared to other Native Americans, so the observed differences between the Aleut and Kamchatkans are to be expected. The magnitude of back migration necessary to obtain the observed frequency of the 9RA seen in the Chukchi and Koryak is 92%, and both the Chukchi and the Koryak are more closely related to the other Native American populations than to the Aleuts. The likely explanation is that the Chukchi and Koryak split from the same ancestral source as the Aleut earlier, followed by genetic separation. Gene flow undoubtedly played a role across all populations throughout the entire process with small movements back and forth, but four major dispersions after the Bering Incubation Model (BIM) best explain the distributions of the ABO haplotypes observed to date.

Based on multidimensional scaling and neighbor joining methods of analysis, three primary Native American clusters and one Siberian cluster from the same ancestral population are shown in Figure 7.8. These results support the Beringian Incubation Model (BIM), which proposes that the ancestral population entered Beringia and then paused because of blockage by glaciers. During this pause, mutations followed by genetic drift resulted in a series of unique markers, such as 01v542 and 9RA.

The three-wave migration theory into the Americas with a standstill in Beringia agrees with previous genetic, linguistic, and archaeological evidence (Reich et al., 2012). The first wave traveled along the west coast, passing through the Tlingit population and then splitting into two South American groups: (a) Parakana and Aymara, Chile, and (b) Mazahua and Aymara of Bolivia. The second wave was inland and ended along the east coast, connecting the Nahua, Maya, Cayapo, Yanomama, Arara, and Chile. The UPGMA tree suggests that the Huilliche because of the highest frequency of

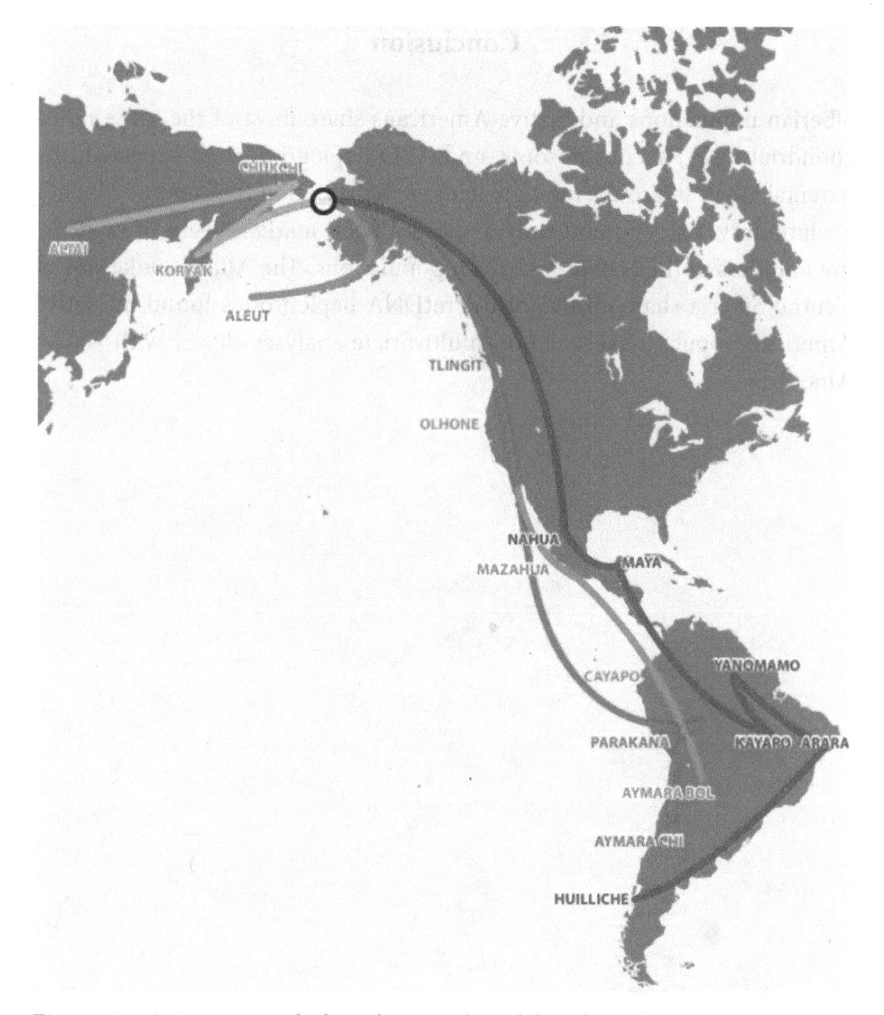

Figure 7.8 Migratory paths based on combined data from the MDS and phylogenetic trees.

Source: Reproduced with permission from Boyd, J. T., *Peopling of the Americas: ABO Blood Group Haplotypes as an Indicator of Native American Origins and Migration from Siberia*. M.A. Thesis, University of Kansas, 2014.

the O1v542 mutation may have split from other Native American groups (see Figure 8.4 in Chapter 8). The third wave into the Americas occurred much later, leading to the formation of an Aleutian group. There is also a suggestion of a fourth split in Meso-America that leads to central South American populations that differ from the coastal populations.

Conclusion

Siberian populations and Native Americans share most of the same mitochondrial DNA, Y-chromosome, and ABO haplogroups. As expected, the parental Siberian groups have greater genetic variation (because of a longer evolutionary history), and only a subset of the markers were brought by the founders of the Native American populations. The Altai populations of Central Siberia share all five of the mtDNA haplogroups found in Native American populations, based on multivariate analyses cluster with Native Americans.

8

Aleutian Islands

Small-Island Evolution

Human populations have experienced a series of fissions and geographic expansions, with a genesis in Africa more than 200,000 years BP, to the far corners of Eurasia, the Americas, and Oceania. The causes and consequences of such migrations depended on ecological, climatic, social, demographic, and biological factors. This chapter focuses on the causes and consequences of Aleut (endonym: Unangan) migrations and settlements from Siberia, across Beringia, and ultimately to an archipelago consisting of 200 islands distributed over 1,800 square kilometers of the Pacific, between North America and Asia. The chronology of this migration was initially reconstructed using a synthesis of archaeological, geological, and ethnological evidence (Jochelson, 1933). The primary objectives of my initial study of the Aleutian Island populations were (1) to determine the genetic relationships among Beringian populations; (2) to test models for the peopling of the Aleutian Archipelago based on molecular genetic evidence from contemporary populations; and (3) to characterize the demographic and genetic sequelae of European contact on the populations of the Aleutian Islands. This research was completed in collaboration with Professor Dennis O'Rourke, University of Utah. This chapter focuses on the causes and consequences of the migrations and settlements of Unangan (Aleut) populations expanding out of Siberia along Beringian landmass to settle in an archipelago consisting of 200 islands distributed over 1,800 kilometers between North America and Asia (Figure 8.1). The chronology of this out-of-Siberia migration was reconstructed based on a combination of archaeological, geological, and molecular genetic evidence (Crawford and West, 2012; West et al., 2010).

In Search of Human Evolution. Michael H. Crawford, Oxford University Press. © Oxford University Press 2024.
DOI: 10.1093/9780197679432.003.0008

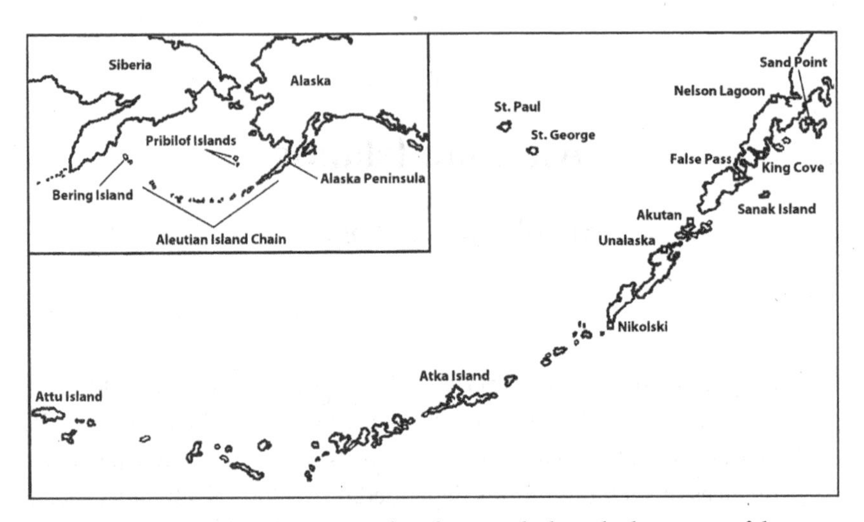

Figure 8.1 Map of the Aleutian Archipelago, including the locations of the Alaska Peninsula, Kamchatka, and islands between Asia and Alaska.

History

The earliest evidence of human habitation in the Aleutian Archipelago is based on stratigraphy and radiocarbon dating (^{14}C dates) of 9,000 to 8,000 years BP from sites located in the eastern Fox Islands at Anagula and Hog Island in Unalaska Bay (Davis and Knecht, 2010). Central islands (Andreanof Islands) of this archipelago have sites dating back to 5,000 years BP. The westernmost islands (Rat Islands, Attu, and Shemya) were settled approximately 3,000 years ago. Apparently, the earliest Aleuts crossed the Bering land bridge more than 9,000 years ago and colonized the islands nearest to the Alaska Peninsula and in a westward direction reaching the Far Islands only 3,000 years ago (West et al., 2007). The prehistoric Aleut populations failed to reach the most westerly islands of the archipelago, Commander Islands (Bering and Mednii), and the northeastern Pribilof Islands (St. George and St. Paul). Russians forcibly relocated Aleut hunters (1825–1830) from Umnak and Unalaska to the Pribilof Islands to harvest fur pelts for commerce in Europe.

What drove these migrations of the Siberians into the Americas? Why was this movement of Arctic hunters primarily in a westerly direction? Why did the Unangan cross Beringia and move in a westerly direction from island to island instead of moving in an easterly direction from Kamchatka?

Population Dynamics

Given the relative scarcity of food and limited technology, most populations of hunters and gatherers tend to consist of extended families of 50–60 individuals (Crawford, 1998). However, in a population migrating into eco-niches with abundant resources, the mortality decreases and fertility increases, resulting in population fission and further migration. There is evidence based on the optimal foraging theory that applies to migrating Aleut populations, which posits that organisms first locate, capture, and consume calorically rich food resources (Winterhalder and Smith, 1981). Aleuts maximized their net energy intake per unit time by colonizing previously uninhabited islands. Evidence of this survival tactic was observed in the earliest site on Shemya Island, where the largest cod was initially caught by the first migrants, then followed by smaller fish and then harder-to-catch birds. When Aleuts depleted their top-ranked resources, they first relied on smaller, harder-to-obtain and less caloric-rich resources. The population increased in size, followed by fission and movement to adjoining islands to acquire previously untapped foraging territories.

Causes of Migration: Volcanic Eruption

There is evidence from the Anagula Blade site (located on Ananiuliak Island) that volcanic eruptions destroyed food sources and forced the Aleuts to migrate and relocate to adjoining islands. Blade tools excavated from the Anagula site were covered by 10–20 cm layers of volcanic ash, most likely associated with a volcanic eruption that created the Okmok Caldera (McCartney and Veltre, 1996). They suggested that the Anagula people fled to the Four Mountains islands to escape this Okmok eruption. Recent volcanic activity in 2008 of the Kasatochi volcano demonstrated the destructive effects of major volcanic eruptions on the surrounding Bering ecosystem and resulted in forced human migration. Events such as migrations and interactions between groups leave a genetic imprint on populations. Lydia Black (1981) has discussed the effects of volcanism on human ecology.

Climatic Changes

Climatic variation played a significant role in Aleut migrations. Savinetsky et al. (2011) analyzed diatoms from natural peat bog deposits on Adak

Island and demonstrated that the Central Aleutian Islands were inhabited during the coldest period of the Holocene. Cold climatic intervals coincided with the highest biodiversity, including the remains of saffron cod, an extremely cold-tolerant fish dating back to 6,000 years BP. This cooler but drier climate was associated with weaker cyclonic activity, less wind, and calmer seas. Another window of opportunity for Aleuts to migrate from the central islands to the western Near Islands of the archipelago occurred approximately 4,000 years ago. Peat bog analyses on Shemya Island revealed a cooler but drier climate, accompanied by weakened cyclonic activity that resulted in less wind and calmer seas. These climatic conditions enabled long-distance interisland travel (e.g., from Rat Island to the Near Islands a distance of 221 km) to previously uninhabited islands in the far west.

Culture Contact and Colonization

The earliest Russian contact with the Aleuts began in the 18th century with voyages of exploration by Vitus Bering and Aleksei Cherikov. The Commander Islands were accidentally "discovered" by Bering and his crew when they were shipwrecked on their return voyage from the exploration of the Americas (Jochelson, 1933). After spending the winter on Bering Island, the survivors returned to Siberia with a harvest of fur pelts that triggered a rush by *promyshlennki* (fur hunters) to the Commander Islands. Between 1824 and 1828, Aleuts from central and western islands were forcibly relocated by the Russian American Company to the Commander Islands to hunt for seals. Because of this contact and relocation, the Aleut population was reduced in size from an estimated 8,000–20,000 in the 17th century to fewer than 2,000 survivors of disease epidemics and warfare (Reedy-Maschner, 2010). Population sizes of Aleuts in the Commander Islands fluctuated from 45 residents in 1825 to a maximum of 626 in 1892. During the 1825–1830 time period, Aleuts, primarily from Unalaska, were forcibly transplanted by the Russian administration to the Pribilof Islands (St. George and St. Paul) to hunt for seal fur.

In 1867, Alaska, including the Aleutian Archipelago, was purchased by the United States from Russia. All of the Aleutian Islands, except for the Commander Islands (Bering and Mednii), were included in the sale and came under US political jurisdiction. This purchase politically isolated the western Aleuts from the central and eastern groups and created a political barrier for any massive population movement

During World War II in 1943, the Japanese Imperial Navy bombed Dutch Harbor in the eastern Aleutian Islands and invaded Attu and Kiska Islands to the west. Aleut inhabitants and US Navy weather monitoring personnel were interred in a prisoner-of-war camp on Hokkaido, Japan. Few of the Attuans survived this internment and on their release returned to the adjoining island, Atka, and to mainland Alaska, leaving Attu unoccupied. The American military also relocated Aleuts living on six islands from the center of the archipelago to evacuation camps in southeast Alaska. As a result, most of the islands in the Aleutian Archipelago are currently uninhabited.

Population Background of Unangans (Aleuts)

The Aleuts primarily subsisted on marine resources, including sea mammals, fish, and various invertebrates. They constructed highly seaworthy kayaks, called "baidarkas," which were ideal for fishing and hunting sea mammals, and capable of crossing ocean expanses between islands. These baidarkas consisted of soft sealskins stretched over a rigid space frame. The frame was usually constructed of driftwood since trees and wood were in scarce supply on the islands. These baidarkas permitted vast migrations that resulted in regional differentiation in linguistic dialects, archaeological assemblages, and regional cultural differences. Dialectic groupings of Unangan language include Eastern Atkan and Attuan (now extinct). As of 2007, only 150 speakers of Aleut remain. Traditionally, the Aleuts lived in subterranean complexes, termed by the Russians as *barbaras*. The Aleuts preferred to call their traditional homes *ulaxin*.

Uniparential DNA Sampling and Analyses

In 1999, under the sponsorship of the National Science Foundation, research assistants from the University of Kansas, Rohina Rubicz and Mark Zlojutro, accompanied by Aleut elder, Alice Petrovelli and I, began an eight-year research program on the populations of the Aleutian Islands and Kamchatka, Siberia (Crawford et al., 2010). Initially, we sampled 267 Aleut volunteers from 11 island populations: Akutan, Atka, Bering (Commander Islands), False Pass, King Cove, Nelson Lagoon, Nikolski, Pribilof Islands (St. George and St. Paul), Sand Point, Umnak, Unalaska, and relocated Aleuts residing in Anchorage. Travel from island to island was by a small single-propeller airplane that flew only in good weather. Buccal swabs and/or blood samples were collected from Aleut volunteers, DNA extracted and analyzed for mtDNA haplotypes and sequences, and NRY chromosome markers (Rubicz et al., 2003; Zlojutro et al., 2009; Crawford et al., 2010). Given the choice between venipuncture and buccal swabs, the vast majority of the US Aleuts were willing to participate and provided buccal swabs. However, Aleuts and other indigenous groups of Siberia preferred venipuncture after providing written informed consent.

Out of the possible mitochondrial DNA haplogroups observed in Siberia and the Americas, only two haplogroups, A and D, are present in contemporary Aleut populations (see Table 8.1). The geographic distribution of these haplogroups in Aleut populations is suggestive of a west-to-east gradient. The statistical relationship between geography and genetics was demonstrated through Mantel tests using correlations between genetic and geographic distance matrices. The relationship between geography and genetics for

Table 8.1 Summary of the Distribution of mtDNA Haplogroups in the Bering Region

Population	N	Haplogroup A	Haplogroup C	Haplogroup D	Other
Unalaska	28	0.607	—	0.393	—
Nikolski	10	0.400	—	0.600	—
Atka	17	0.294	—	0.706	—
St. George	28	0.179	—	0.821	—
St. Paul	35	0.286	—	0.714	—
Bering	35	0.000	—	1.000	—
Akutan	16	0.563	—	0.438	—
False Pass	11	0.727	—	0.273	—
King Cove	33	0.515	—	0.485	—
Nelson Lagoon	16	0.563	—	0.438	—
Sand Point	38	0.447	—	0.553	—
Aleut: Total	267	0.378	0.000	0.622	0.000
Alaskan Yupik	25	0.960	0.000	0.000	0.400
Siberian Yupik	90	0.700	0.022	0.278	0.000
Chukchi	72	0.708	0.097	0.111	0.083
Athapaskan	21	0.952	0.048	0.000	0.000
Koryak	147	0.054	0.361	0.014	0.571
Itelmen	46	0.065	0.130	0.000	0.804

Source: Reproduced with permission from Rubicz, R. C., *Evolutionary Consequences of Recently Founded Aleut Communities in the Commander and Pribilof Islands.* Ph.D. diss., University of Kansas, 2007.

the Aleutian populations was highly significant with $r = 0.72$ and $p < 0.000$ (Crawford, 2007c). In part, this geographic gradient was further enhanced by the relocation to Bering Island by the Russian administrators of specific families bearing only the D haplogroups. The A7 haplotype (A2a1A) observed in Aleut populations differs from NaDene and Eskimo A3 by a specific Aleut mutation 16212 G (Zlojutro et al., 2006). Raff et al. (2010) found the B2 mtDNA haplogroup in 25% of the skeletal remains from the Brooks River site (Raff et al., 2010). These were transitions at np 16189C and 16217C diagnostic for the B2 lineage and suggestive of greater genetic diversity prior to population fission. The presence of B2, A2a1a, and D2a1a in DNA ancient samples was followed by temporal migration and loss of B2 throughout the archipelago. One result of the fission and population migration was the loss of genetic diversity throughout the Aleutian Islands.

The Aleut population of Bering Island contains a single haplogroup D. Fixation of D and loss of A have been attributed to genocide and epidemics by Derbeneva et al. (2002). Given that the frequency of haplogroup A is 30% in Aleut populations, it is unlikely that individuals with A were selectively eliminated. Genetic drift hypothesis demonstrated that fixation of D could not have occurred in six generations, given such a high frequency of haplogroup D in the founding population.

Genetic Evidence of Kin Migration

Based on hypervariable region I and II mitochondrial DNA sequences, the spatial autocorrelation statistical method of Bertorelle and Barbujani (1995) was utilized to further elucidate the relationship between geography and genetic markers. In Figure 8.2, the x-axis contains the "lag geographic distances (kilometers)," while the y-axis displays "product moment coefficients" analogous to Moran's "I." As predicted by the isolation-by-distance model, those populations closest geographically have the greatest, significant correlations. Those populations beyond approximately 1,300 km "lag geographic distance," however, have negative correlations, suggestive of kin migration following population fission and resettlement. Spatial autocorrelation plots are indicative of kin/family migration from island to island and indicate a rapid genetic differentiation of island populations and greater interisland genetic differentiation. The application of population genetic structure to complex diseases is discussed in Crawford (2015).

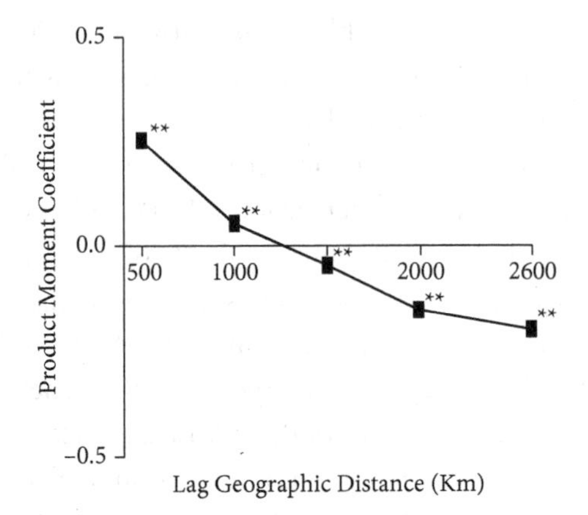

Figure 8.2 Spatial autocorrelation plot of populations from the Alaskan archipelago using mtDNA sequences and the methods of Bertorelle and Barbujani (1995).

Source: Adapted with permission from Crawford, M. H. et al., "Origins of Aleuts and Genetic Structure of Populations of the Archipelago," *Human Biology* 82(5–6): 710. © 2010, The Authors.

In contrast to mtDNA haplotypes found in contemporary Aleutian Islands that are entirely Native American, Y-chromosome haplotypes reveal the extensive gene flow from the Russian military and administrators into the Aleut gene pool. Aleut males were separated from females by being moved to other islands to hunt fur seals. The Russian administration encouraged unions between Aleut females and Russian males as a way of politically controlling the admixed populations. These asymmetric gene pools prompted Reedy-Maschner (2010) to write an article posing the question: "Where did all the Aleut males go?"

Figure 8.2 plots the spatial patterns of mtDNA sequence diversity. The ordinate contains the autocorrelation indices, while the abscissa is the distance in kilometers between the islands. Using the method of Bertorelle and Barbujani (1995), the relationship between geographic distance and mtDNA sequences is highly significant ($p < 0.000$). This plot supports a kin structured migration model—that is, fission occurred along familial lines. The highest correlation between populations is in the 500-km distance, followed by lower correlations in the 1,000-km lag and negative correlations in the greater distances. The relationship between geography and genetics appears to be linear and highly significant statistically, indicating kin migration.

Figure 8.4 presents a plot of the Aleutian Islands based on Monmonier's (1973) Maximum Difference algorithm, which identifies triangulation

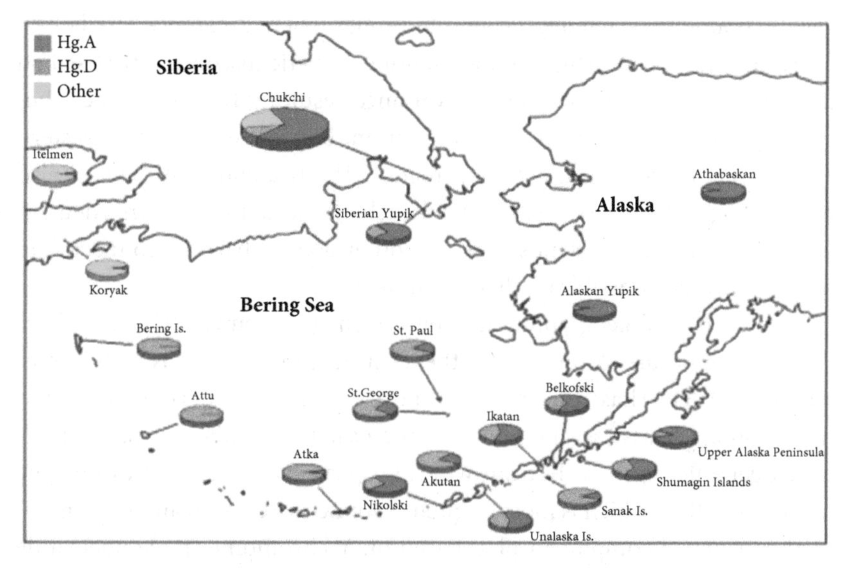

Figure 8.3 Mitochondrial DNA haplotypes distributed over the Aleutian Archipelago, Alaska, and Siberia.

Source: Adapted with permission from Crawford, M. H. et al., "Origins of Aleuts and Genetic Structure of Populations of the Archipelago," *Human Biology* 82(5–6): 701, fig. 3. © 2010, The Authors.

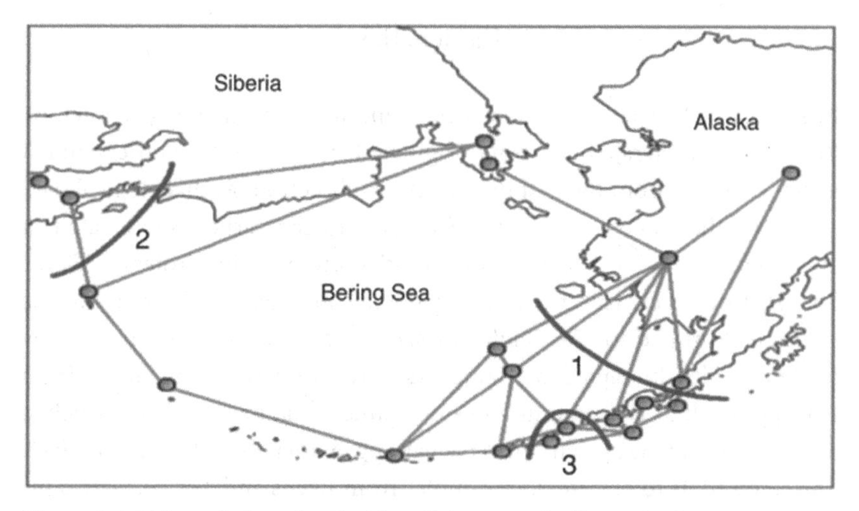

Figure 8.4 Triangulation plot for identifying genetic discontinuity.

Source: Adapted with permission from Crawford, M. H. et al., "Origins of Aleuts and Genetic Structure of Populations of the Archipelago," *Human Biology* 82 (5–6): 708, fig. 8. © 2010, The Authors.

genetic boundaries, namely, geographic zones where differences are largest (Barrier, version 2.2). This method identifies genetic discontinuity based on the proportion of the total genetic variance resulting from differences between groups. In regard to Aleut migration, barriers exist on the western Alaskan peninsula and the central region. The triangulation method also reveals a genetic barrier between Kamchatka and the eastern Aleutian Islands. The Aleutian Islands have two other genetic barriers connected to migrations made possible by changes in weather.

Y-chromosome geographic distribution in the Aleutian Islands reflects Russian admixture primarily in the central and western islands. Thus, haplogroup R1a (East European) is most common on Bering Island and the Pribilof Islands (see Figure 8.5). The eastern Aleutian Islands experienced gene flow primarily from west European fishermen and exhibit the haplogroup R1b, which is most frequently found in Scotland and England.

The principal component plot, based on Y-chromome single nucleotide polymorphisms (SNPs), reveals a clustering of the Aleut populations except for Akutan and False Pass. This variation is most likely due to small sample sizes collected from those islands. St. George and St. Paul cluster with Russian populations due to high frequency of gene flow. Nelson Lagoon and Sand Point reflect west European admixture.

Kamchatka

In August 2001, we wanted to test the hypothesis (originally generated by Russian researchers) that Aleuts originated in Kamchatka and expanded into the islands from the east in a westward direction. Archaeological data suggested that Aleuts came from Siberia with the earliest site located in an eastern island. We set up a collaborative project with Professor Victor Spitsyn from the Institute of Anthropology in Moscow. Spitsyn and his research team had to cross eight Russian time zones to fly from Moscow to Petropavlovsk, capital of Kamchatka. The research team drove from Petropavlovsk to two communities on Kamchatka, namely Esso, which is 520 km north of the capital, and Anavgai, approximately 100 km east of Esso. This allowed us to obtain DNA samples from Evens and Koryaks—two indigenous Siberian populations that had settled in Kamchatka. Following the research in Anavgai and Esso, we returned to Petropavlovsk and then flew on a small airplane to Nikolskoye, Bering Island. Bering Island and Mednii

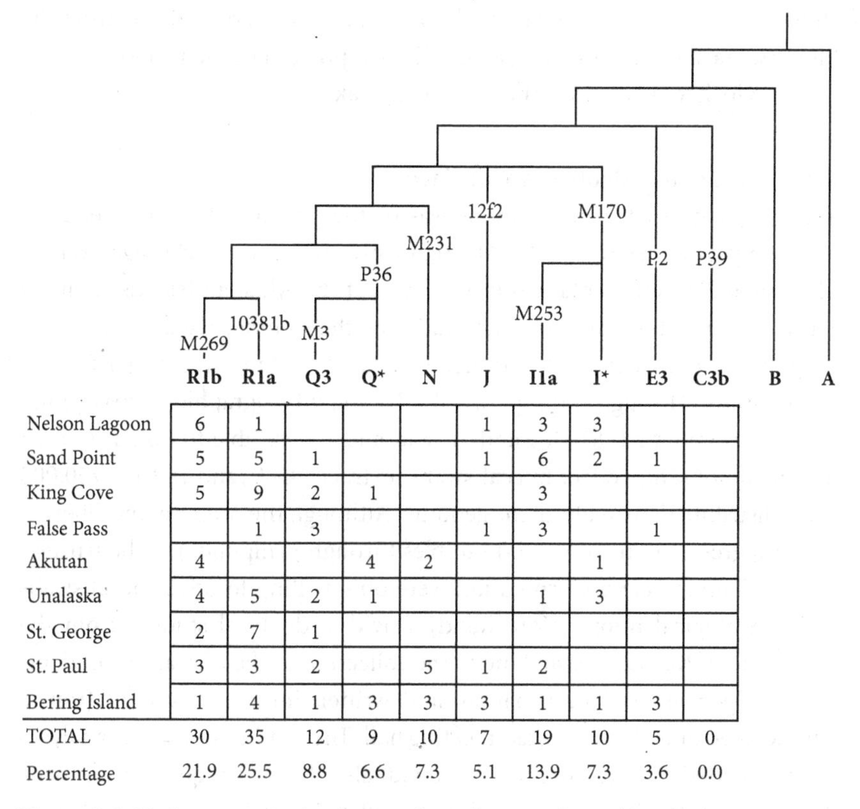

	R1b	R1a	Q3	Q*	N	J	I1a	I*	E3	C3b
Nelson Lagoon	6	1				1	3	3		
Sand Point	5	5	1			1	6	2	1	
King Cove	5	9	2	1			3			
False Pass		1	3			1	3		1	
Akutan	4			4	2			1		
Unalaska	4	5	2	1			1	3		
St. George	2	7	1							
St. Paul	3	3	2		5	1	2			
Bering Island	1	4	1	3	3	3	1	1	3	
TOTAL	30	35	12	9	10	7	19	10	5	0
Percentage	21.9	25.5	8.8	6.6	7.3	5.1	13.9	7.3	3.6	0.0

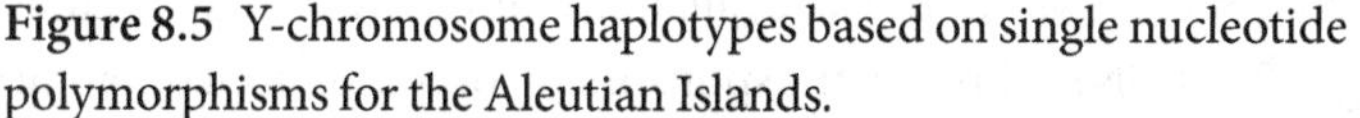

Figure 8.5 Y-chromosome haplotypes based on single nucleotide polymorphisms for the Aleutian Islands.

Source: Reproduced with permission from Zlojutro, M., *Mitochondrial DNA and Y-chromosome Variation of Eastern Aleut Populations: Implications for the Genetic Structure and Peopling of the Aleutian Archipelago*. Ph.D. diss., University of Kansas, 2008.

(Copper) Island are part of the Commander Islands. Although Mednii once had a small Aleut community, it was vacated during the 1960s. Because of severe fog, bad weather, and fracturing my left tibula in an accidental fall on the side of a volcano, we were one week late returning to Petropavlovsk. This created a crisis because we would overstay by a few days our agreement with the Russian National Academy of Sciences. At that time a single commercial flight of 1,500 miles was available per week from Petropavlosk to Anchorage. The "apparatchiks" (Russian contemptible administrators) of Kamchatka demanded written permission from the Academy of Science, located in Moscow, for an extension of our stay for a few extra days. However, academician Spitsyn was able to persuade the Kamchatkan administrators to allow us to stay a few extra days without formal written permission from the Moscow-based Academy of Science. The US portion of the team flew from Petropavlosk to Anchorage the following week.

Resampling of the Aleutian Archipelago

The sampling of Aleut populations was further expanded in October 2015 during a yearly meeting of the Aleutian Corporation in Anchorage, Alaska. More than 200 representatives from eleven of the Aleutian Islands attended this meeting. After I gave a lecture outlining the significance of the research program to the assembled trustees, 116 Aleuts volunteered to participate in this study. Through the support of a National Geographic Society grant, GENO 2.0 Grant #178163, Randy David and I were able to obtain written informed consent, collect buccal swabs, extract DNA, and test for 750,000 SNPs distributed throughout the genome. Although the Aleuts of the Siberian Islands agreed to provide blood samples through venipuncture, the trustees of the Aleutian Island Corporation were only willing to offer buccal swabs and genealogical information. Randy David, a doctoral student from the University of Kansas, assisted me in the collection of the buccal swabs, demographic/genealogical information, and written informed consent from a table located outside the trustee-meeting hall. The results of the DNA analyses were provided (confidentially) to individuals who had requested feedback during the initial interviews. With the availability of GENO 2.0 SNP chips characterizing 750,000 SNPs through the sponsorship of National Geographic Society (NGS), this survey of the Aleutian Islands was much more informative than the earlier analyses based only on uniparental markers mtDNA and NRY. Gene by Gene Limited Laboratories genotyped 145 Aleuts using an Illumina platform and the newly developed NGS SNP chip.

Results

Varying degrees of admixture with Russian, English, and Scottish immigrants were recorded, depending on geographic location, of individuals sampled. Table 8.2 summarizes Russian ancestry for eight geographic zones with the highest Russian/European admixture in the Pribilof Islands and the Eastern Archipelago. There was low Aleut ancestry, only 7.6%, in the few individuals who did not provide their place of birth to the researchers. On average, 51.4% of the genes in the Aleutian Islands were of European ancestry.

Table 8.2 Admixture Estimates of Ethnic Aleuts Residing in Different Regions of Alaska and Siberia

Population	Aleut	Other (European/Russian)
Bering	59.7	40.3
Eastern Peninsula	56.8	43.2
Pribilof Islands	38.2	61.8
Central Archipelago	48.1	51.9
Eastern Archipelago	40.0	60.0
Mainland Alaska	48.9	51.1
Mainland United States	40.5	59.5
Unknown, POB	7.6	92.4
Total	48.6	51.4

Principal Component Analyses

Principal components analyses (PCA) is a useful method for evaluating ancestry since the genes mirror the evolutionary and demographic history of human populations. PCA is a dimension reduction method, and in this study it was performed on 170,636 SNPs from the 750,000 SNPs initially characterized using the GENO 2.0 chip (Patterson et al., 2006). The SNPs were pruned using the computer program PLINK, which removed SNPs that were in linkage disequilibrium with other SNPs (Purcell et al., 2007). Individuals missing more than 10% of their SNPs and in cases of relatives with IBD values of 0.2, one of the two SNPs was removed from the analysis. PCA captures variation in multidimensional data and reorganizes it along orthogonal axes, ordered by the variances that they capture (Price et al., 2006). The 170,636 SNPs and 2,499 individuals (72 remaining from the Aleuts) were aggregated and provided a glimpse into the genetic structure and ancestry of Aleuts, by comparing them to Africans, East Asians, Europeans, Native Americans, and South Asians compiled from the 1000 Genome Project (The 1000 Genomes Project Consortium, 2015). The first PC axis separated the Africans from all other populations, while the second PC separated East Asians and Europeans. A few of the Aleuts clustered among Europeans indicating considerable Russian or west European admixture. In the PCA, individual Aleut participants were distributed between East Asians and European populations with several culturally identified Aleuts clustering solely with European genomes.

Admixture Analysis

Depending on methods used and genetic markers employed, there is considerable variation as to the estimated proportion of Aleut versus Russian/European ancestry. If uniparental markers (mtDNA) are utilized for diallelic estimates, 100% of Aleuts with known Native American ancestry display only indigenous American haplogroups—"A" and "D." However, only 15% of Aleuts displayed Native American NRY markers. Admixture estimates based on autosomal STR markers suggest that 60% of Aleut ancestry-informing sites are indigenous American and 40% are the result of gene flow from European sources (see Figure 8.5). Admixture estimates for Bering Island (the least admixed of all islands in the archipelago) based on 750,000 SNPs provide similar ancestry proportions of 60% Aleut and 40% Russian gene flow (see Table 8.3).

The Admixture 1.3 program was utilized for estimating ancestry in a model-based analysis from a large autosomal SNP data set using unrelated individuals (Alexander et al., 2009). The "K-means" algorithm sorted individuals into an increasing number of clusters and yielded a Bayesian Information Criterion (BIC) value for K. The K value with the lowest BIC yields evidence of the best fit for the data. Equal sample-sized populations of genomes were used in this analysis to represent six geographic regions (downloaded from the 1000 Genomes Project) and compared to the Aleuts (Crawford et al., 2021). The best fit for the admixture analysis (with the lowest cross validation error BIC = 1414.64) was $k = 2$ suggestive of

Table 8.3 Admixture/Ancestry Estimates Based on Uniparental Markers, Autosomal Short Tandem Repeats, and Genomic 750,000 Single Nucleotide Polymorphisms

Genetic Marker(s)	Percentage Aleut (Indigenous American)	Percentage Russian/European
Mitochondrial DNA	100	0
Nonrecombining Y STRs	15	85
Autosomal STRs	60	40
Genomic SNPs	60	40

SNPs: single nucleotide polymorphisms; STRs: short tandem repeats.

Source: Reproduced with permission from Zlojutro, M., *Mitochondrial DNA and Y-chromosome Variation of Eastern Aleut Populations: Implications for the Genetic Structure and Peopling of the Aleutian Archipelago*. Ph.D. diss., University of Kansas, 2008.

Aleut and European (Russian/Scandinavian ancestry). The next best fit (BIC = 1417.395), $k = 3$ reflected an earlier evolutionary event, the replacement of Paleo-Aleuts by Neo-Aleuts (see Figure 8.6).

Since 1999, we have conducted research on the origins and genetic structure of indigenous populations of the Aleutian Islands primarily based on uniparental markers—mtDNA and NRY. These studies revealed the following: (1) The underlying Aleut genetic structure was preserved in the maternal genomes with an exceptionally high correlation ($r = 0.68$, $p > 0.004$) between geographic and genetic distances among 11 islands distributed from the Alaska peninsula to Kamchatka, Siberia. (2) In contrast to mtDNA markers, no significant correlation was observed between NRY markers and geography. (3) Only 15% of indigenous Aleut Y-chromosomes were observed in Aleut males. Most of the NYR markers originated in Russia, Scandinavia, or England.

This is a follow-up study using buccal swabs from 115 volunteer Aleuts, attending a corporation meeting in Anchorage, Alaska, in 2014. An additional 30 blood samples were available from Aleuts of Bering Island from

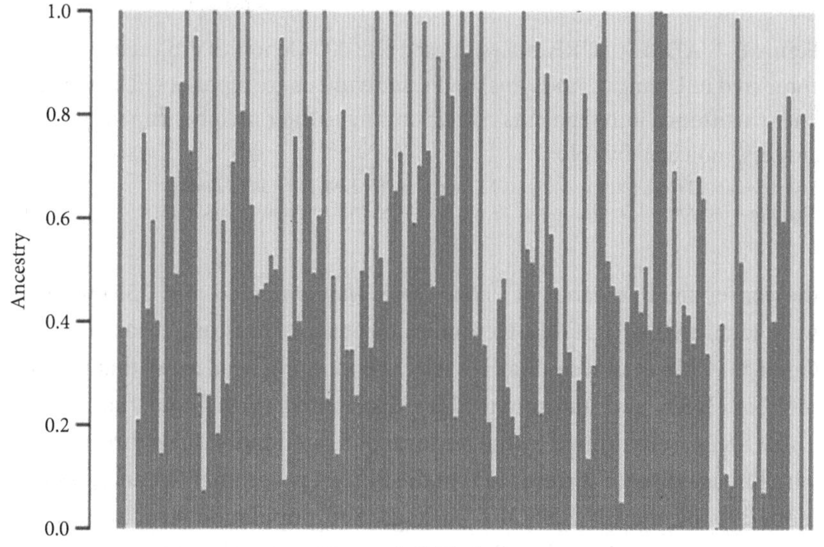

Figure 8.6 Population structure inferred from an ADMIXTURE analysis. ADMIXTURE indicates two clusters ($k = 2$). Each individual is represented by a vertical stacked column indicating the proportions of ancestry in k constructed ancestral groups.

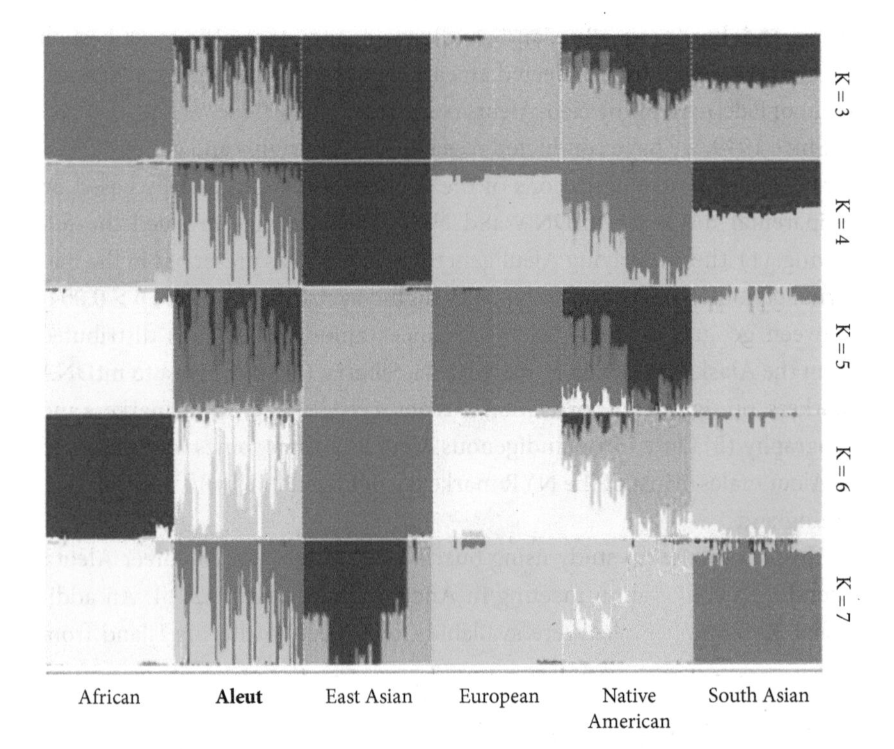

Figure 8.7 ADMIXTURE analysis based on 1000 Genomes Project data and compared to Unangax autosomal SNPs. Individuals are represented by vertical lines partitioned into segments corresponding to their membership in genetic clusters, indicated by color.

Source: Adapted from Muñoz-Moreno, M., and Crawford, M. H. (Eds.), *Human Migration: Biocultural Perspectives*, p. 29, fig. 2.6. © Oxford University Press 2021.

the earlier field research of 2004. DNA was extracted and 750,000 SNPs were identified by FTDNA Genomics Center of Houston, Texas. This new Geno 2.0 Chip is an illumina HD select genotyping bead array that includes mtDNA, NRY, and autosomal SNPs distributed throughout the genome. These SNPs were pruned by the removal of related individuals through pair-wise linkage disequilibrium (LD) in the PLINK program. PCA analyses were plotted using the EIGENSOFT package, and population structure was revealed using the ADMIXTURE Program. Intra- and interpopulation diversity were estimated using Arlequin software version 3.5.

ADMIXTURE 1.3 program was used for estimating ancestry in a model-based analysis using a large autosomal SNP genotype data set with unrelated individuals (Alexander et al., 2009). K-means algorithm sorted individuals

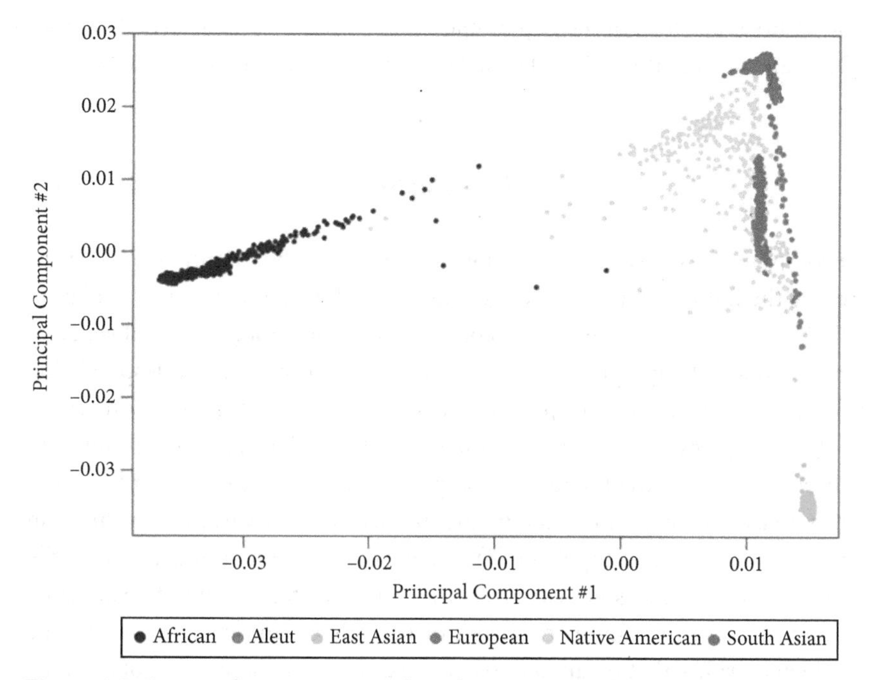

Figure 8.8 Principal component analysis for six ancestral populations (k). The ancestral populations include Africa, Aleut, East Asia, Europe, Native America, and South Asia.

Source: Adapted from Muñoz-Moreno, M., and Crawford, M. H. (Eds.), *Human Migration: Biocultural Perspectives*, p. 28, fig. 2.5. © Oxford University Press 2021.

into an increasing number of clusters and then provided a Bayesian Information Criterion (BIC) value of K. The K with the lowest BIC is the best fit for the data. Using unsupervised clustering (adegenet), the best fit for the number of gene clusters in the Aleut dataset is K = 1 with the Bering Island sample being the least admixed population.

Mitochondrial DNA sequences preserved the genetic structure of Aleutian populations, as evidenced by the high correlation of genetic distances with geographic distances. Residents with Aleut mothers exhibit either the A or D haplotypes. Gene flow from Russian males to Aleut females is reflected in only 15% of the Y-chromosome haplogroups Q or Q3—Native American. Remaining haplogroups are of European (E, I, I1a) or Eurasian (J, N, R1a, and R1b) origin. This pattern of gene flow (Russian males and Aleut females) resulted from a policy advocated by the Russian governor of Bering.

The use of 750,000 mutational markers across the entire genome provided greater precision and statistical power when compared to population structure reconstruction using uniparental markers. Comparative data from NGS allowed identification of gene flow from specific regions of Europe.

Conclusion

Evolutionary consequences of the settlement of the Aleutian Islands include the following: (1) Loss of genetic variability resulted from population fission and founder effect. This conclusion is confirmed by the absence of mitochondrial haplogroup "B" in contemporary populations, despite being found at 25% in a small prehistoric site. (2) Spatial autocorrelation results reflect frequent kin migration. This form of familial migration resulted in increased genetic drift, and hence rapid genetic differentiation of the more distant island populations, and further reduced genetic variation in Unangan populations of the archipelago. (3) Genetic discontinuity is illustrated by the presence of mating/genetic barriers that reflect historical climatic variation. Long-distance migrations occurred principally during periods of extreme cold and aridity that resulted in pacific seas. By contrast, periods of cyclonic activity and tempestuous seas coincided with relatively less population movement and greater genetic differentiation of island group populations. (4) Mitochondrial DNA sequences preserved interisland maternal population structure and migration patterns. mtDNA evidence supports the high correlation of genetic distance with geographic distance ($r = 0.72$; $p < 0.000$). This intimate relationship between genetics and geography was observed in contemporary populations despite numerous sociopolitical and demographic upheavals, such as World War II relocation to Anchorage and the 1825 forced relocation of Unangan males to the Commander and Pribilof Islands for the harvesting of seal furs. (5) No statistically significant correlation was observed between genomics and geography based on NRY genetic markers. This finding is due to the clear pattern of gene flow between Russian *male* migrants and Unangan *females*. Only 15% of the contemporary Unangan male population exhibited indigenous American NRY haplogroups, Q and Q3. Remaining haplogroups are of European (E, I, I1A) and Eurasian (J, N, R1a, and R1b) origin. Post-contact gene flow into the Aleutian Islands was from Russian males in the central and western regions; English and Scandinavian fishermen entered the Eastern Islands after the sale of Alaska by Russia.

9

Bio-demography of Italian and Hungarian Populations

Introduction

The genetic structure of human populations can be analyzed using demographic data. Under some political and social situations, it is not possible to obtain genetic data from specific populations, even though their genetic structure may provide significant evolutionary information. Normally, demographic and social data are gathered together with the genetic and morphological data by teams of researchers in the field (Crawford et al., 2007d). During 50+ years of field investigations, in several studies it was not always possible to collect genetic data due either to the reluctance of the communities or the specifications of the supporting granting agencies. I had to rely on bio-demographic analyses, based on church records, documents, and family interviews, to assess affinities among populations based on migration matrices. In this chapter, I will review two such examples: (1) a study of three alpine villages experiencing a breakdown of reproductive isolation in northern Italy and (2) 24 agricultural communities isolated by a bend in the Tisza River and affected by a change in the location of the Russian border (Crawford et al., 1999). The Tiszahat project was supported by the Earthwatch Foundation and staffed by volunteers experiencing fieldwork for the first time while receiving class training in genetics and anthropology.

Italian Valley—Valle Maira

My research on human population dynamics was initiated during the summer of 1968. A colleague from University of Pittsburgh, Professor Alan McPherron, who had conducted archaeological excavations in Yugoslavia, was a strong advocate for expanding anthropological genetic research to this geographic region. When I attended an academic meeting at the former

In Search of Human Evolution. Michael H. Crawford, Oxford University Press. © Oxford University Press 2024.
DOI: 10.1093/9780197679432.003.0009

dacha of Generalisimo Tito (located in Andrejevlhe), little was known about the genetics of Serbian populations. Yugoslavia was just emerging out of Soviet domination, and as a result there were numerous restrictions on foreign research. I met most of the Serbian and Croatian anthropologists at this meeting but became aware of the difficulties of conducting genetic research in Yugoslavia. I returned by train with a stopover in Torino, where I visited Professor Brunetto Chiarelli, the head of the Institute of Anthropology at the University of Torino. He convinced me that the alpine valleys south of Torino offered better opportunities for evolutionary research of isolated villages.

Valle Maira is an Alpine valley of northern Italy, carved by the meanderings of the Po River with an entrance approximately 40 km northwest of the City of Cuneo and 100 km south of the city of Torino (see Figure 9.1). This valley is 43 km in length, narrow, with villages lining a single road traversing its length. Some of the villages located off the main road are inaccessible from four to six months of the year due to heavy snowfall and danger of avalanches during the winter months.

According to the 1962 census, the total population of Valle Maira was 15,472 inhabitants, with the village of Dronero at the mouth of the valley being most populous—that is, consisting of 6,670 inhabitants. Administratively, the valley was divided into nine communes, which are comparable to townships. This study focused on three communes: Acceglio (with 482 inhabitants located at the eastern terminus of the valley), Elva (with 94 inhabitants and geographically the most isolated community, north

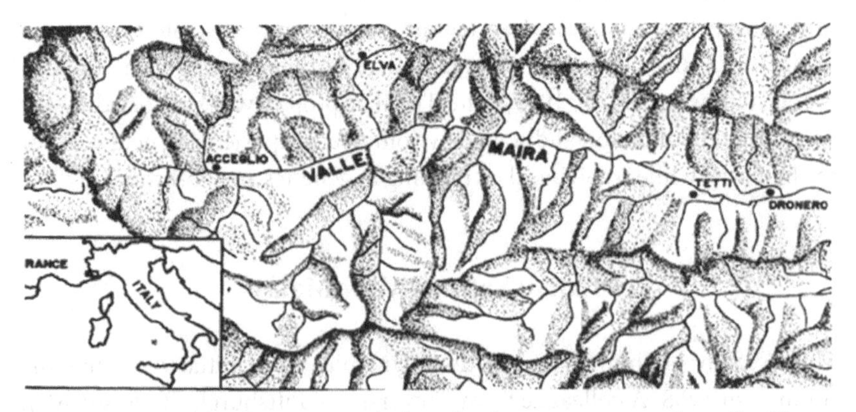

Figure 9.1 Map of Valle Maira, including the three communes, Acceglio, Elca, and Dronero, sampled in this study.

Source: Reproduced with permission from A. W. Eriksson et al. (Eds.), *Population Structure and Genetic Disorders*. © 1980, Elsevier.

Figure 9.2 Photograph of Valle Maira with the village of Acceglio visible at a distance. This village is closest to the French border to the west of the commune.

of the Valley), and Tetti, a hamlet (*frazione*) of the commune of Dronero, located at the mouth of the valley. These three communities were selected for study to examine varying degrees of geographic isolation and potential gene flow (see Figure 9.2).

Participants

A research team from the Anthropological Institute of the University of Torino consisted of graduate students, including Chiara Bullo, Anna Vidili, Lorenzo Grosspietro, and Anna Imasi. We spent three months conducting research on three villages: Acceglio (endpoint of the valley near the border of France was divided into seven hamlets), Dronero (village closest to the entrance to Valle Maira), and Elva (the most isolated village north of the valley). This research team focused on transcribing church records to reconstruct changes in migration, mating patterns, and marital distances over a period of several generations. Standard demographic questionnaires were administered to the heads of 260 households from the three communities. These forms were designed to reconstruct genealogies and to obtain demographic information on three or four generations. However, we learned that the residents of the three villages did not wish to participate in genetic aspects of the study, particularly involving the drawing of blood through venipuncture. At that time, buccal swabs were not an option, and DNA extraction was too expensive and technically not feasible.

The church records from Valle Maira offered an opportunity to measure the breakdown of reproductive isolation in a series of genetic isolates. I defined "genetic isolate" as small populations with varying degrees of reproductive isolation (Crawford, 1980). This study attempted to document the degree of reproductive isolation of alpine populations and recent increase in immigration and emigration resulting from tourism and industrialization during the last 50 years in large urban centers, such as Torino and Milano, proximal to the valley. The demographic analyses documenting the gradual disintegration of social isolation included changes in the frequencies of isonymous marriages (marriages between mating pairs having the same surname and possibly sharing common ancestry) and geographic expansion of marital distances (North and Crawford, 1996). In small village populations with a limited number of surnames, frequency of isonymy provides a rough estimate of isolation and inbreeding. Lasker (1985: 525) observed that "Surnames are used by human biologists because in many cultures they are inherited like genes." Graf et al. (2010) demonstrated the relationship between surname distributions and Y-chromosome markers in Aleut populations.

Demographic Analyses

An examination of the incidence of exogamous marriages, over four generations, reveals that Elva was reproductively the most isolated with only 4% of marriages in the earliest generation being exogamous (see Table 9.1). The commune of Acceglio displayed the next most frequent exogamy rates and displayed the disintegration of population isolation. During the last generation, more than one of every two marriages came from outside the village. This breakdown of reproductive isolation came from migration to larger towns, Torino and Cuneo, for employment and tourism into the valley.

The commune of Acceglio had been declining numerically during the last 50 years. The census of 1951 revealed a reduction of 47% from 1951 to 1967. This drastic reduction in population size was a result of two world wars, which constricted the population pyramid through a combination of

Table 9.1 Frequencies of Exogamous Marriages in Three Italian Alpine Communities

Generation	Dates	Acceglio	Elva	Tetti
4	1890–1910	7.91	4.00	10.00
3	1911–1930	4.32	6.15	15.25
2	1931–1950	20.00	15.45	32.00
1	1951–1968	50.49	25.00	52.00

Source: Reproduced with permission from A. W. Eriksson et al. (Eds.), *Population Structure and Genetic Disorders*. © 1980, Elsevier.

factors: rapid increase in mortality of young men killed during the wars and reduction in fertility plus high westward emigration to France and to the industrial cities of the north and south.

Similarly, the incidence of individuals with the same surname marrying, that is, isonymy, is highest in the most isolated community, Elva (see Table 9.2). In small isolated communities with a limited number of surnames, isonymy provides an approximation of degree of inbreeding (Lasker, 1985). Only four surnames were present in the church records of Elva, and this community displays from 21% to 12% isonymous marriages. Tetti, the least isolated *frazione*, displays from 9% to 2% isonymous marriages. Analysis of Tarpa (northeastern community) based on isonymy and repeated pairs (RP) revealed the population structure of this Valle Maira community from 1780 to 1979 (Duggirala et al., 1992).

The degree of reproductive isolation was assessed on the basis of marital exogamy and constancy of surnames in the population over time. Genetic isolation underwent a rapid change during World War I, Benito Mussolini's

Table 9.2 Frequencies of Isonymous Marriages during Four Time Periods in Three Italian Alpine Communities

Generation	Dates	Acceglio	Elva	Tetti
4	1869–1888	15.12	21.50	9.00
3	1889–1908	15.00	16.31	10.26
2	1909–1928	3.90	12.32	4.32
1	1929–1948	4.13	12.60	2.10

Source: Reproduced with permission from A. W. Eriksson et al. (Eds.), *Population Structure and Genetic Disorders*. © 1980, Elsevier.

ascent to power, and World War II. Geographic barriers such as mountains influence mate selection and marital movement. However, economic factors such as employment in heavy industry attract laborers from the valley into cities at a disproportionate rate.

Conclusion

Although genetic documentation of the breakdown of reproductive isolation, based on blood groups and protein markers, was not possible in the 1960s, currently there is greater interest in molecular genetics and individual ancestry. It may now be possible to document the genetic changes associated with population movement.

Tiszahat, Hungary—Subdivided Agricultural Populations

Introduction

In 1985, the National Academy of Sciences funded a pilot study by Michael Crawford conducted in Tiszahat. In 1986, Earthwatch Foundation funded an expedition to Tiszahat, Hungary, for 15 volunteer/participants to obtain training and unique cultural experiences. A team of faculty and graduate students of anthropology from Kosuth University at Debrecen, plus several US and United Kingdom professors were organized to provide lectures and training and to supervise research by the volunteers. Native Hungarian graduate students from Kosuth University were paired with the English-speaking volunteers to record church records and to interview Tiszahat families.

Participants

Two Hungarian faculty members, Professors Miklos Pap and Katalin Szilagyi, from the Department of Evolutionary Zoology and Human Biology at Kosuth University, obtained the necessary governmental permissions for this research program and helped train the volunteers. In addition, 10 graduate students from Kosuth University were paired with the Earthwatch volunteers.

Professor Tibor Koertvelyessy, from Ohio University, a Hungarian emigre to the United States during the Hungarian uprising of the 1960s was a bilingual Hungarian/English speaker with considerable field experience and knowledge of Hungarian history. He lectured to the Earthwatch volunteers about the use of church records for bio-demographic studies.

Dr. Rosalind Harding, currently an Associate Professor of Biological Anthropology at Oxford University, also provided lectures on the genetic structure of human populations based on demographic data. Her Ph.D. dissertation at LaTrobe University of Melbourne focused on demographic aspects of population biology.

Professor Peter Nute, Department of Anthropology, University of Washington, provided lectures on genetic theory in the training sessions for the volunteers.

History

Tiszahat is an agricultural region of northeastern Hungary outlined by a bend in the Tisza River and the Russian border. Tiszahat has been highly stable numerically with settlements dating back to the 12th century AD. However, in 1945 after World War II, a new Hungarian-USSR Russian border was imposed, partitioning the region. In addition, migration across the Soviet border was difficult due to political reasons. Tiszahat populations were chosen for study because of their subdivision into 21 villages and geographic isolation from the remainder of Hungary. The Tisza River inundated the region periodically, creating a swampy land infested with mosquitos and malaria. Because of its geographic isolation and disease patterns, Tiszahat remained culturally and genetically homogeneous and avoided invasion by various European and Asian armies. However, despite the cultural homogeneity, a religious dichotomy into Catholic and Calvinist congregations was observed in several villages of Tiszahat. The demographic consequences and possible evolutionary implications of the Tiszahat region split into a religious dichotomy has been earlier discussed by Koertvelyessy et al. (1992, 1993).

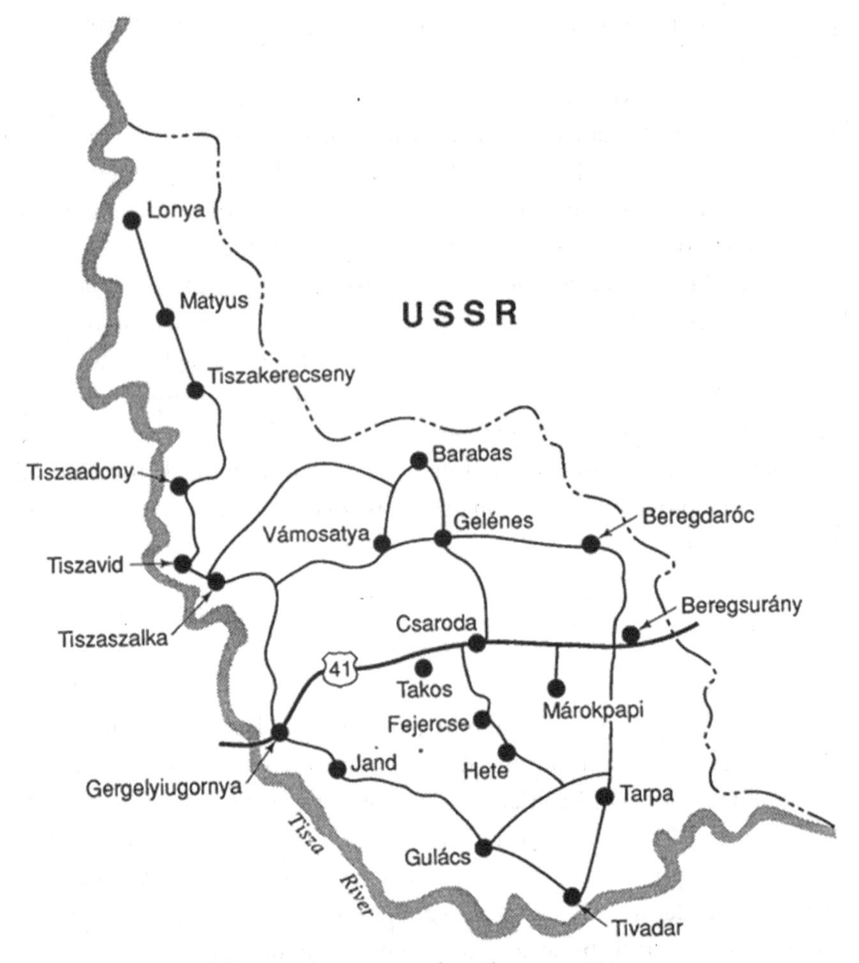

Figure 9.3 Map of the Tiszahat region of northeastern Hungary.

Source: Reproduced with permission from Crawford, M. H. et al., "The Effects of a New Political Border on the Migration Patterns and Predicted Kinship (PHI) in a Subdivided Hungarian Agricultural Population: Tiszahat," *Homo-journal of Comparative Human Biology* 50(3): 201–210. © 1999, Elsevier.

Methodology

Calvinist Church records were transcribed for all 24 of the Tiszahat villages from the end of the 18th century to 1986. Some of these villages in close geography proximity had merged because of their numerical growth, and the records had to be combined. As a result, matrimonial migration patterns were measured for 20 villages (Crawford et al., 1999). Migration patterns for two time periods, 1875–1899 and 1950–1974, were compared. These periods of time were chosen to contrast patterns of migration prior to World War I and II versus migration after the establishment of a new border and the political subdivision of Tiszahat. Surname repetition and isonymy were compared as a measure of population structure in Tiszahat communities (Koertvelyessy et al., 1990).

Two marital migration matrices were constructed with male and female villages of birth represented by rows and columns. Based on these matrices, predicted kinship $Ø_{ij}$, a measure of the genetic similarity between populations, was computed (Morton, 1973). The diagonal elements of the matrix provide the probability that two random genes in a population were identical by descent and denoted local kinship (Crawford et al., 1999). The off diagonal elements of the matrices measure the genetic relationship between any two populations: i and j. High values denote close population affinity, while low values suggest little genetic resemblance (Crawford et al., 1999).

Results

A comparison of predicted kinship Phi between two time periods reveals a significant reduction in Phi from 1875 to 1979. Specific comparisons of pairs of villages reveal a breakdown in isolation in some villages and no statistically significant differences in others. Only two villages (Gergelyiugornya and Beregsurany) out of 20 comparisons had a slight increase in predicted kinship. Given the increase in migration and breakdown of isolation from 1875 to 1979, it is not surprising that the likelihood of random genes being of common descent is reduced.

The correlation between the Phi values for all 20 communities during the two time periods is 0.56, which is highly significant ($p < 0.001$). A matrix correlation between Phi and geography (as measured by trail distances between the villages) exhibits a negative association of $r = -0.58$ NS. As shown in Figure 9.4, the relationship between Phi and geography is curvilinear for both time periods, resulting in a pseudo-leptokurtic curve (Crawford et al., 1999).

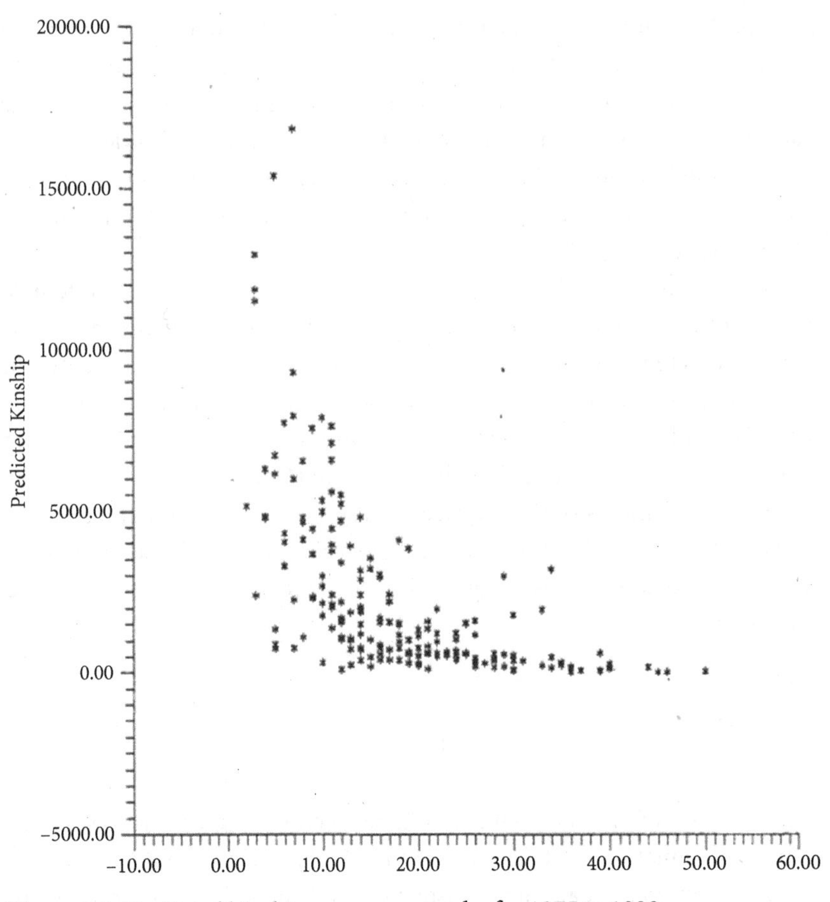

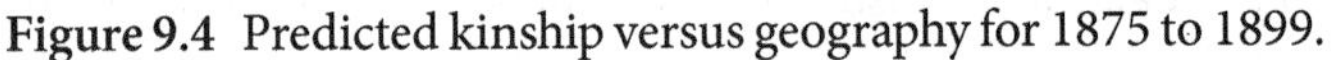

Figure 9.4 Predicted kinship versus geography for 1875 to 1899.

Source: Reproduced with permission from Crawford, M. H. et al., "The Effects of a New Political Border on the Migration Patterns and Predicted Kinship (PHI) in a Subdivided Hungarian Agricultural Population: Tiszahat," *Homo-journal of Comparative Human Biology* 50(3): 201–210. © 1999, Elsevier.

Conclusion

The bio-demographic results generate hypotheses that can be tested using genetic data. In the Tiszahat study the relationship between geography and predicted kinship decreased over the two periods and resembled Malecott's (1969) isolation-by-distance model. The effects of the imposition of a new border between Hungary and the USSR in 1945 resulted in a reduction of endogamy and a decrease in predicted kinship. The closing of the border in 1945 created a linear effect as a result of greater gene exchanges between adjacent villages and the elimination of gene flow from an easterly direction. Application of molecular genetic data to test the demographic results would permit the testing of hypotheses generated by the church record–based migration data. "Tiszahat population structure has been affected in a single generation by a unique historical effect"—that is, the creation of a new border (Crawford et al., 1999: 209).

10

Genetic Epidemiology

Primate (Baboon) Models of Disease

Sukhumi Baboons and Outbreak of Leukemia/Lymphoma

In 1964, I visited the Sukhumi Primate Center in Abkhasia with hopes of obtaining blood samples from the massive *Papio hamadryas* (baboon) colony for characterizing the genetic structure of captive colonies (see Figure 10.1). Unfortunately, its director, Boris Lapin (a decorated and wounded pilot from World War II), was not present during my initial visit, but the assistant director escorted me on a tour of the facilities. I was informed at that time that it was not possible to ship blood samples from the USSR to the United States. Later, while on a National Academy of Sciences Exchange to the Soviet Union in 1976, I revisited the Sukhumi Primate Center again and discovered that an outbreak of malignant lymphoma (malignancy of lymphoid tissue) was sweeping through the baboon colony (Lapin, 1975). Boris Lapin (director) asked me to develop a collaborative research program on the causes of this apparent epidemic. According to Lapin, in the 1960s Soviet scientists inoculated 12 baboons with cells from hospitalized human leukemia patients (cancer of leukemic cells that make up bone marrow and blood) causing infection and the eventual death of 300 baboons from the 3,219 animals housed in the open-air enclosures of Sukhumi Primate Center (Voevodin et al., 1996). The death rate from lymphoma was considerable at approximately 12 baboons per year. Human T-cell lymphotropic virus type 1 (HTLV-1), a related virus, was implicated as the etiological agent.

A number of significant questions arose: (1) Was the malignant lymphoma caused by a virus introduced from human hosts? (2) Once infected, was the infectious agent transmitted from baboon to baboon through physical contact such as bites? (3) Since the lymphoma clustered among specific baboon sibships, did genetic predisposition play any role in lymphoma transmission?

In Search of Human Evolution. Michael H. Crawford, Oxford University Press. © Oxford University Press 2024.
DOI: 10.1093/9780197679432.003.0010

Figure 10.1 Free-ranging baboons of Sukhumi, located in a colony on the outskirts of the City of Sukhumi. No evidence of malignant lymphoma in this colony of baboons.
Source: Photo courtesy of Dennis O'Rourke.

(4) Why did the free-ranging baboons in a separate colony on the outskirts of Sukhumi show no signs of the lymphoma?

Because of an oncology agreement between the US National Cancer Institute (NCI) and the USSR Academy of Medical Sciences, I was able to initiate a research program on the baboon colony and obtained blood samples and pedigrees. The inbreeding coefficient for 1,226 members of the colony, reconstructed using baboon pedigrees, was a modest F = 0.027. However, with the removal of unrelated baboons (i.e., F = 0.0) from the analysis, the inbreeding coefficient increased to 10% (0.096) higher than any F-value documented for human populations. The highest known inbreeding coefficient in human populations was F = 0.04. The inbreeding coefficient (F) is the probability that two genes at any locus in an individual are identical by descent. There was no relationship between increased risk of contracting lymphoma and level of inbreeding (Crawford et al., 1984). In the 1970s, whole genomic DNA sequencing was difficult from a technical and financial standpoint; however, the baboon specimens were characterized genetically by an assortment of protein markers. The only significant association, at 0.005, was between phosphoglucomutase loci, PGM*1 and PGM*2, and the incidence of lymphoma infection. A statistical association between PGM loci with high anti-VCA-HVP titers was demonstrated in the baboon colony (Crawford and O'Rourke, 1978). Based on path analyses of baboon sibships, approximately 55% of the phenotypic variance (death from lymphoma) is explained by direct parent-to-offspring transmission.

Human T-cell lymphotropic virus type 1 (HTLV-1) was implicated by researchers from NCI as the etiological agent of the Sukhumi baboon lymphoma (Blattner et al., 1982). These were the only genetic analyses possible at that time prior to the molecular revolution and the sequencing of the viral DNA.

Surprise!

In 1996, Voevodin and colleagues observed on the basis of the molecular similarity between baboon and macaque viruses that a possible interspecies transmission had occurred between Rhesus macaque STLV-1 (simian T cell leukemia/lymphoma virus type) to the Sukhumi baboons (Schatzl et al., 1993). Partial *env* gene sequences of all four STLV-1 isolates from Sukhumi lymphomatous baboons were 97% to 100% similar to the sequence of known Rhesus STLV-1 (Voevodin et al., 1996). They concluded that on the bases of 37 Sukhumi STLV-1 isolates that interspecies transmission occurred and caused the outbreak of lymphoma in Sukhumi baboons.

Two Ph.D. dissertations by graduate students from the Laboratory of Biological Anthropology at the University of Kansas examined the underlying genetic bases of morphological structures in free-ranging baboons. Dennis O'Rourke measured the body of each baboon that was anesthetized by a dart and capture gun, for purposes of venipuncture. He examined the relationship of "Papiometrics" to the degree of inbreeding (F). At the same time, Robert Baume used amalgam to make dental impressions and casts for each baboon and examined the relationship of odontometrics and dental morphology to the inbreeding coefficient (Baume and Crawford, 1978, 1980).

Conclusion

In answer to the original questions posed:

1. Virus (STLV-1) appears to be the etiological agent of the Sukhumi baboon lymphoma outbreak.
2. Judging from the spatial distribution of infections in baboon sibships, the virus is most likely transmitted through bites, thus inoculating other animals.
3. It is not clear whether the occurrence of the lymphoma infection is influenced by a genetic predisposition or that related baboons are more likely to socially interact and inoculate each other.
4. Free-ranging baboons in the geographically separate colony did not exhibit any lymphoma because most likely they were not exposed to interspecific contact with the macaques. What is more puzzling is how and when did the macaques interact with the baboons? The apparent lymphoma outbreak was initially traced by the Russian scientists to the inoculation of the baboons with human leukemia cells.

Using field research on animal models can provide considerable insight into genetic–environmental interactions in complex diseases such as leukemia or lymphoma.

11
Basque Origins and Genetic Structure

Introduction

This research on the genetic structure and origins of the Basque people of northern Spain and western France was based on a collaborative project between researchers from a number of different laboratories, institutions, and intellectual approaches. In March 2000, the Laboratory of Biological Anthropology received DNA from Dr. Antonio Arnaiz-Villena (Universidad Complutense de Madrid) a small sample of 20 unrelated persons from Vizcaya Province in Spain (Arnaiz-Villena et al., 2002). We compared their genetic affinities to other European populations by testing for 13 tetrameric short tandem repeat (STR) loci. The allelic frequencies of these polymorphic STRs were compared with 21 European and North African populations for nine of the typed STRs (Zlojutro et al., 2006). Neighbor-joining trees and multidimensional scaling plots of D_A genetic distances revealed that the Vizcayan Basques are an outlier to neighboring Iberian and North African populations. This result raised the question about the origins of the Basques: Where did they come from?

Dr. Arantza Apraiz-Gonzalez, a Spanish/Basque, postdoctoral fellow in the Laboratory of Biological Anthropology (LBA) at the University of Kansas, followed up on the analyses of the Vizcaya Basque samples and initiated a field research program on the origins of the Basques. She received her Ph.D. in anthropology from the University of the Basque Country, Bilbao, Spain, under the mentorship of Professor Rosario Calderon, currently at the University of Madrid. Dr. Apraiz was co-investigator of a grant proposal on the origin of the Basques submitted to the National Geographic Society. This grant received excellent reviews and was awarded to the LBA at the University of Kansas (Project 6935-00). Being of Basque/Spanish ancestry helped Dr. Apraiz-Gonzalez gain access to a total of 35 mountain villages (in the western Pyrenees) and recruit 652 autochthonous participants (who claimed four Basque grandparents). Buccal swab DNA was obtained from populations residing in four Spanish Basque provinces (Alava, Vizcaya, Guipuzkoa, and Navarro). These field investigations in Spain were conducted by a single researcher during a 2000 to

In Search of Human Evolution. Michael H. Crawford, Oxford University Press. © Oxford University Press 2024.
DOI: 10.1093/9780197679432.003.0011

2002 timeframe (see Figure 11.1). The DNA from village Basques constitute a unique sample, much more informative than earlier pilot studies of urban Basque schoolchildren and admixed Basque/Spanish adult populations from cities such as Bilbao. DNA from Basque buccal swabs was extracted by Kristin L. Young (research assistant at the LBA) employing standard phenol-chloroform protocols (Young et al., 2012). Mitochondrial DNA haplogroup sites were identified through the analyses of restriction fragment length polymorphisms (RFLPs) plus the sequencing of the first hypervariable segment of the control region (HVS-1) Young et al., 2012.

Kristin Young analyzed the Basque DNA and wrote her dissertation "The Basques in the Genetic Landscape of Europe" at the University of Kansas (Young, 2009). Because of her training in both analytical methods and molecular sciences, she collaborated on this Basque project with Dr. Apraiz and me. She identified uniparental haplogroups for mtDNA and supervised mtDNA sequencing at the University of Kansas. She analyzed mtDNA

Figure 11.1 Basque provinces sampled by Dr. Arantza Gonzelez Apraiz.

Source: Adapted with permission from Young, K., *The Basques in the Genetic Landscape of Europe*. Ph.D. diss., University of Kansas, 2009.

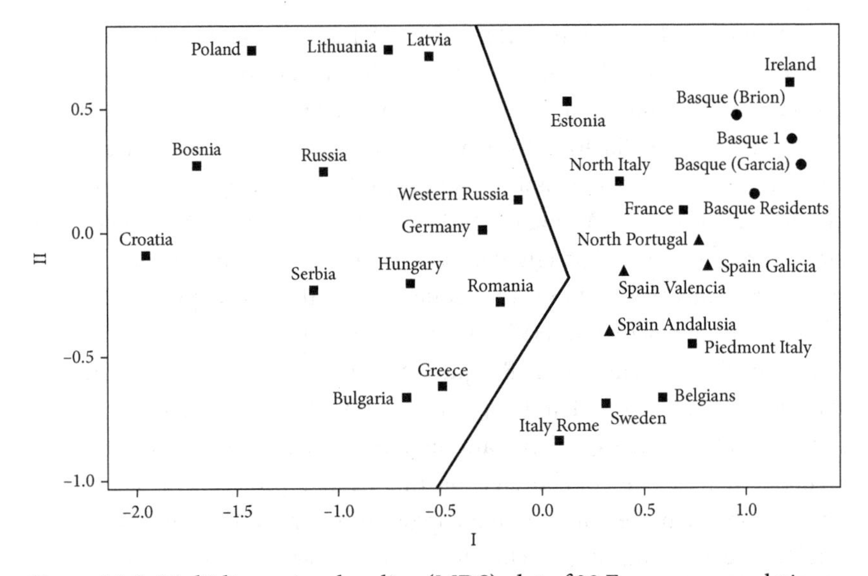

Figure 11.2 Multidimensional scaling (MDS) plot of 29 European populations using Y-chromosome markers.

Source: Adapted with permission from Young, K., *The Basques in the Genetic Landscape of Europe*. Ph.D. diss., University of Kansas, 2009.

and Y-chromosome data and, using the molecular results, tested several hypotheses concerning the origins of the Basque populations. Since the LBA was not set up for STR analyses at that time, that research was supervised by Professor Ranjan Deka in his laboratory at the University of Cincinnati Medical Center. Dr. Guangyun Sun analyzed the Y-STR haplotypes in Dr. Deka's laboratory (Young et al., 2011a, 2011b).

History

The origins of Basque populations have fascinated anthropologists, linguists, and geneticists for many decades. For a summary of the genetic history of the Basques and their geographic location, see Calafell and Bertranpetit (1994). Earlier, Ammerman and Cavalli-Sforza (1984) after reviewing the known blood marker frequencies for European populations concluded that the contemporary genetic variation in Europe is the result of Neolithic farmers expanding out the Middle East. This demic diffusion model (DDM) attributes the linguistic spread of Indo-European languages to bands of Neolithic farmers. Novelletto (2007) based on Y-chromosome variation has argued that the expansion of farmers into Europe 10,000 years ago carried their technology but not necessarily genes. This introduction of technology without gene flow into Europe has been termed the cultural diffusion model (CDM). The Basques are one of a few, non-Indo-European speaking populations of Europe. They speak a language, Euskara, considered a linguistic isolate, unrelated to any other extant language in Europe. Indo-European languages were likely introduced into Europe by technologically advanced Neolithic farming populations from the Middle East. The Basques do not speak any of the Indo-European languages, much like the linguistic isolates—Hungarian and the Lapp of Finland.

Origin Hypotheses

Three major theories about the origins of the Basques have appeared in the literature:

1. Basques share recent common ancestry with populations of the Caucasus. This theory was based primarily on human leuckocyte antigens (HLAs) and immunoglobulin analyses (GMs) Bertorelle and Barbujani, 1995.
2. Basques are descendants of ancient Iberian populations who migrated to the peninsula from North Africa following a major climatic change in the Sahara between 4000 and 8000 BC (Arnaiz-Villena et al., 2002).
3. They are a remnant population and descendants of Paleolithic Europeans, who evolved in Spain with little gene flow from the Neolithic farmers (Calafell and Bertranpetit, 1994).

Results

Genetically, the Basques are distinct from other European populations for several classical markers, namely, cde Rhesus blood group (>50%), GM*ZAG (35%), AcP*B (73%), and ADA*1 (>97%). Based on PCA analyses of R matrix using 13 standard loci and 21 alleles, the Basques are separated from all other groups.

On a molecular basis, Basques have high frequencies of uniparental haplogroups likely to be Paleolithic in origin: mtDNA and NYR. These haplogroups include Y-chromosome R1b and mtDNA common European haplogroups H and U5. HVS-1 sequences place the date of population expansion to the Paleolithic, arguing in opposition to the complete replacement DDM of the Neolithic transition.

Table 11.1 Mitochondrial DNA Haplogroups Identified in Basque Populations of Spain

Province	H	I/X	J	N1b	T	V	W	U/K	U	K	Not Determined	Total
Alava	71	0	3	0	2	5	3	1	21	4	8	118
Vizcaya	111	2	20	0	3	15	1	4	31	10	36	233
Guipuzcoa	102	1	6	1	1	14	1	10	27	10	52	225
Total	314	3	29	1	6	34	5	15	80	24	104	615

Note: Total includes samples from a small number of individuals from Navarre.

Source: Reproduced with permission from Young, K. L., Devor, E. J., and Crawford M. H., "Demic Expansion or Cultural Diffusion: Migration and Basque Origin," in *Causes and Consequences of Human Migration*, pp. 224–249. © 2012, Cambridge University Press.

Young et al. (2011a) examined the genetic structure of 27 mountain villages ($n = 377$ individuals) from contemporary Basque populations based on nine autosomal short tandem repeat loci (D3S1358, D5S818, D7S820, D8S1179, D13S317, Di8S51, D21S11, FGA, and vWA). The multidimensional MDS plot of Shriver's DSW distance matrix revealed that the Basques differed from both the populations of the Caucasus and North Africa. Thus, STR analyses do not support the theory that the Basques shared common ancestry with these populations.

An MDS plot of 31 populations using nine autosomal STR loci reveals that the Basque populations are separated from all other groups along the first axis. These molecular results parallel blood group and protein based European variation.

Conclusion

Basques are a European ethnic group that experienced considerable genetic drift and some gene flow since the last Glacial Maximum. Although Basques and the people of the Caucasus speak non-European, agglutinative languages (such as Hungarian), a close genetic relationship between them is not supported by DNA evidence. Genetic barriers were detected and no common, non-European ancestor was suggested. Thus, Basques are a remnant European population that has experienced some admixture, particularly with geographically adjacent Iberian populations.

Based on mtDNA haplogroups, Y-chromosome paternal markers, and autosomal STRs, no statistically significant genetic heterogeneity was observed in locus-by-locus analysis of molecular variance (AMOVA) comparisons of autosomal STRs between the Basque provinces. HVS-1 sequences suggest that the date of Basque population expansion was in the Paleolithic, arguing against a complete replacement demic diffusion of a Neolithic transition.

12
Conclusion

Introduction

The topical and geographic breadth of 50+ years of my field investigations are revealed in Table 12.1. This table summarizes my professional career as an anthropological geneticist. I organized and participated in more than 13 major projects, and this table lists the local sponsors and funding agencies that supported my research. Some of these projects were relatively short-lived, conducted by postdocs and graduate students; I played a minor role, mainly obtaining funding and permissions, and writing sections of the publications. While on a Fulbright fellowship, I collected blood samples based on caste status at the University of Andhra Pradesh in India. DNA analyses and subsequent publications comparing the castes of India were generated by former University of Kansas (KU) graduate student Michael Bamshad (MD) and University of Utah Professor of Genetics Lynn Jorde (Bamshad et al., 1996). In the fishing villages (out ports) of Newfoundland, the collection of the blood samples and genetic marker analyses were carried out by the Canadian Red Cross Blood Center research assistant Marilyn Collins, under the direction of Dr. Richard Huntsman and postdoctoral fellow Dr. Tibor Koertvelyessy. My role in this research was data analysis and publication of the results. The study of the farming communities of Tiszahat, Hungary, was part of an Earthwatch summer training and research program, and no blood samples or DNA was collected.

This book provides the reader with a glimpse of the methodological history of anthropological genetics, by presenting a series of chapters roughly in chronological order of the fieldwork and data collection. The first field expeditions during the 1960s and 1970s preceded the molecular and the informational revolutions; thus, no DNA analyses were possible. Populations were characterized genetically through the observed variation in red blood cell groups, serum proteins, and erythrocyte proteins. Whole blood samples, taken by venipuncture, were collected into vacutainers containing preservative (ACD) and shipped, packed in ice, to laboratories in the United

In Search of Human Evolution. Michael H. Crawford, Oxford University Press. © Oxford University Press 2024.
DOI: 10.1093/9780197679432.003.0012

Table 12.1 Field Research Programs Conducted by the Laboratory of Biological Anthropology (LBA) from 1968 to Present (Crawford, 2007d)

Populations	Chronology	Sites	Local Sponsors	Funding Agencies
Italian Alpine	1968	Valle Maira	U. of Torino	U. of Pittsburgh
Tlaxcaltecans	1969–1975	City of Tlaxcala, San Pablo del Monte, Saltillo, Cuanalan	Ministry of Health	NIH
Irish Itinerants	1970	Dublin, Wexford	Medical Social Board of Ireland	Wenner-Gren Foundation, University Center U. of Pittsburgh
Baboons of Sukhumi	1974–1980	Institute of Experimental Pathology & Therapy, Sukhumi, Abkhasia	Russian Academy of Medical Sciences	NCI (contract)
				NIH, KU, Biomedical Research
Black Caribs (Garifuna)	1975–1982	Belize, Guatemala,	Ministry of Health	NIH, NSF
	2010–2018	St. Vincent, Honduras	Medical School in Dominica & Honduras	
		Roatan, Dominica		
Castes of India	1990–1991	Andhra Pradesh	University of Andhra Pradesh	Fulbright (CIES)
Yupik and Inupik Eskimos of Alaska	1976–1977	St. Lawrence Isl., Wales, King Island	Norton Sound Health Corporation	NSF
Fishing Outports	1981–1983	Newfoundland	Canadian Red Cross	Wenner-Gren Foundation
Mennonites	1979–2018	Kansas and Nebraska	General Conference and Holderman Churches	NIA, Attorney General's Settlement Fund Earthwatch,
Farming Villages of Hungary	1986–1988	Tiszahat, Hungary	Debrecen U.	
	1989–2001			
Siberia	1999–2005	Central Siberia, Kamchatka	Soviet Academy of Science,	National Academy of Science
	2010–2011		Academy of Medical Science	NSF

(*continued*)

Table 12.1 Continued

Populations	Chronology	Sites	Local Sponsors	Funding Agencies
Unangan (Aleuts)	1999–2006	Aleutian Archipelago	Aleut Corp	NSF
			Pribloff Island Association	
			Tribal Councils	
Basque villages	2001–2003	Northern Spain	Village elders were contacted by Basque postdoc Dr. Gonzalez-Apraiz	National Geographic Society Grant (6935-00)

NCI, National Cancer Institute; NIA, National Institute of Aging; NIH, National Institutes of Health; NSF, National Science Foundation.

States for analyses. After the blood samples were genotyped in a laboratory, the resulting data were key punched for computer analysis. These data were primarily analyzed using computer programs at the Computer Center, University of Kansas. I remember how thrilled the members of the LBA were when we received funding from the National Science Foundation to install a computer terminal connected to the university main frame. This installation of the terminal eliminated the numerous trips to the Computer Center necessary to collect and view the results of the research.

Team Research

Interdisciplinary team collaboration can accomplish significantly more research in a shorter time period than projects generated and carried out by a sole investigator. Several of my research programs would not have been possible without the participation of cultural and biological anthropologists, archaeologists, and physicians. The Tlaxcala research design would not have been developed without input from cultural anthropologist and colleague Professor Hugo Nutini. He had invested several decades of research in the Tlaxcala-Puebla Valley and reconstructed the ethnohistory of the populations. His knowledge and insight provided a time dimension to the transplantation of the Tlaxcaltecans from the Valley of Tlaxcala to the Valley of Mexico in 1525. The massive transplantation of 400 families from the Valley of Tlaxcala to Saltillo in the 1570s provided another time dimension and marked a significant fission of the Tlaxcaltecan gene pool. This unique interaction of Cortes, the Spanish Crown, with the Tlaxcaltecans provided historical insight into the Spanish governmental policies.

I learned about the unique history of the Black Caribs/Garifuna from cultural/applied anthropologist Nancie Gonzalez. We conducted two joint field investigations in Livingston, Guatemala, and in several communities of Belize. In addition, historical and social background provided by Gonzalez allowed the interpretation of the genetic and biological results.

Field investigations on the genetics of biological aging in Mennonite populations would not have been possible without the participation of Professor John Janzen. Janzen's family was part of the immigration of the Mennonites from Crimea to Kansas in the 1860s. John Janzen is a highly respected member of the Goessel community, and his participation made the fieldwork that included 25 collaborating researchers possible.

Why Field Research?

Crawford (2007d), in a chapter on the significance of field research in anthropological genetics, reviewed the following: (1) Importance of adding a comparative dimension to a study or analysis. Complex diseases, such as hypertension or diabetes, are influenced by the action of unique environments and provide insights into genetic–environmental interactions. (2) The addition of a time dimension of several hundred years to the research design. In the case of the Tlaxcaltecans, a time dimension was included in the study through the historical reconstruction of population fission. Similarly, the history of the populations provided the study with examples of population fission and the possibility of testing hypotheses concerning the genetic and morphological differentiation and population structure. (3) Unique social and demographic structures of populations. Field research provides access to genetic isolates, such as the Mennonites of central Kansas, and the analyses of complex phenotypes, such as biological aging (Crawford, 2000, 2005).

Why Do Communities Participate?

Many members of indigenous communities are curious about their origins and history. For example, the Aleuts of Alaska are fascinated by their origins, specifically, where did they originate in Siberia, who migrated, and when? The Aleut populations were willing to participate in studies of their history, but they were less interested in disease etiology. The Aleuts felt that earlier they were deceived by physicians in historical/cultural studies, but under the guise of disease etiology.

Ethical Considerations

The Nuremberg Code, written in response to the Nazi human experimentation, provided an original set of ethical standards for research on human individuals. The code requires (1) voluntary consent; (2) the disclosure of the nature, duration, and purpose of the experimentation; (3) description of the methods employed; (4) revelation of any potential hazards of the research—either physical or social; and (5) discussion of any possible benefits or lack of direct benefits to society.

The American Anthropological Association's (AAA) Principles of Professional Responsibility provided the following ethical field research guidelines:

1. Do No Harm

The AAA Code of Ethics states: "Anthropological researchers must do everything in their power to ensure that their research does not harm the safety, dignity or privacy of the people with whom they work." A book written by journalist Patrick Tierney, *Darkness in El Dorado: How Scientists and Journalists Devastated the Amazon*, charges prominent scholars with unethical misbehavior in their work with an Amazonian tribe. Tierney claimed that Napoleon Chagnon violated professional ethics for 30 years by encouraging conflict and falsifying evidence to exaggerate the ferocity of the Yanomami. Tierney accused Chagnon with ethically dubious field methods such as bribing children for information about tattoos, taking photos without permission, and improper involvement in conflicts between villages. Tierney also accused James Neel, an MD geneticist, of deliberately injecting tribal people with a controversial measles vaccine (Edmonston B vaccine), causing a massive epidemic. The American Society of Human Genetics investigated the charges and found them to be gross misrepresentations and basically false.

The AAA appointed a committee to determine if further action should be taken. This committee recommended the creation of a five-person task force to inquire into Tierney's allegations. Since the AAA does not accredit or issue credentials to its members, it cannot investigate, censure, or expel anyone for ethical misconduct. The task force's final report was posted on the AAA website in May 2002. It was a mixture of opinions, interviews, meditations on ethics, and recommendations.

Each researcher should consider and discuss possible ways that the research might cause harm to members of the population. The AAA warns of possible damage, such as harm to dignity, and bodily and material well-being. Potential unintended consequences of the research and its impact on individuals, communities, and environments should be considered. For example, in biological anthropology and genetics, potential risk associated with venipuncture, such as infection or the accidental creation of hematomas, must be revealed to potential research subjects. To date, the

American Association of Anthropological Genetics has not yet published specific guidelines for ethical principles of field investigations (Turner et al., 2018). Ethical guidelines have been proposed by the American Association of Physical (Biological) Anthropologists (2003) and the American Anthropological Association (1971). Turner (2005) edited the proceedings of a symposium held in 2001 at an AAPA meeting. This edited volume covered a cross-section of the field of biological anthropology and included chapters on genetics and human biology. Two chapters discuss research on ancient DNA and focus on how agreements were reached with different indigenous communities (O'Rourke, 1980).

2. Be Open and Honest Regarding Your Work

Researchers should be clear and open regarding the purposes, methods, outcomes, and sponsors of the research. Ethical requirements include openness, honesty, transparency, and informed consent. Since 1974, Congress has enacted federal regulations requiring the approval of the methods of research by an Institutional Review Board (IRB), who assure the rights and welfare of participating subjects. Generally, IRB approval is required for studies in the fields of health, genetics, and social sciences. They protect subjects from potential physical, psychological, or social harm.

The Havasupai tribe sued scientists at the Arizona State University (ASU) for misrepresenting a genetic study as a medical project. The tribal lawsuit alleged that researchers from ASU used DNA for studies without proper consent. The Havasupai were awarded several million dollars in damages. Politically charged interpretations of this dispute have fanned tribal distrust of academics.

3. Obtain Informed Consent and Necessary Permissions

Obtaining permission varies depending on the country, the background of the participants, and the nature of the proposed research. If the research includes a health component, permissions should be obtained from the community, chief health officer, and/or the minister of health. In the case of research on St. Vincent Island, the chief medical officer granted permission for our research. These communities and populations were contacted prior to the commencement of the research. The research methodology and risk factors must be fully explained and permission obtained. It is wise to include local physicians or scientists from the community.

Informed consent is a process for obtaining permission before conducting research on a subject. Informed consent is obtained either in written form or orally. If oral informed consent is requested, a witness should be present who can confirm that all regulations have been followed. The following components must be present for valid informed consent: (1) purpose of the study—that is, what is to be accomplished?; (2) any discomfort or risk must be explained; (3) benefit of the project to the participant or society; (4) the subject should have the freedom to participate or to quit the study without prejudice; (5) alternative procedures, if possible, should be described (Crawford, 2007). For example, DNA can be obtained from an individual either through venipuncture or by buccal swabs, which are less invasive.

4. Competing Ethical Obligations

Anthropologists must weigh competing ethical obligations to research participants, students, professional colleagues, and funding agencies. In anthropological genetics, methodologies should be considered with the least amount of risk versus the quantity and quality of data. As an example, Siberian indigenous populations agreed to the use of venipuncture for DNA studies. In contrast, Alaskan Aleuts were only willing to provide buccal swabs with the understanding that the DNA will only be used to explore their origins in Siberia but not for biomedical research.

5. Make Your Results Accessible

Feedback on the results of the study should be provided to members of the community as quickly as possible and disseminated in clear, understandable ways. The Aleut project provided three different methods of informing the participants of the results of the research: (1) I addressed the yearly meeting/dinner of the island trustees and village representatives. Several hundred Aleuts attended this meeting in Anchorage and, after my lecture, asked many probing and revealing questions. (2) The Aleut Corporation publishes a monthly newsletter, *Aleut Current*, that I was invited to contribute to. It is widely read by Aleuts from all of the islands. (3) Publications in scientific journals and books describing the results of the research were made available to the Aleut Association and Corporation. I compiled a special issue of the journal *Human Biology*, devoted entirely to Aleut genetics, demography, history, linguistics, and archaeology (Crawford et al., 2010). The Aleutian Corporation and Pribilof Island Association received copies of this special journal issue and distributed it among interested members of the community.

The Mennonites of the Midwest were most interested in the genetics of biological aging. The communities had several physicians in residence and preferred that the results of the individual blood chemistries should be transmitted to their primary care physicians for possible insight into the patient's current health.

6. Protect and Preserve Records

As stated by the AAA Administration in 2012: "Anthropologists have an ethical responsibility for ensuring the integrity, preservation, and protection of their work." I protected the confidentiality and security of the demographic and primary DNA data. There was considerable interest in the community about individual mitochondrial and single nucleotide polymorphism (SNP) results. We provided individual phenotypes to those requesting their own results. However, no individual data were shared with family members or unrelated members of the community, anxious to determine potential paternal relationships. DNA samples were frozen and kept in the Laboratory of Biological Anthropology at the University of Kansas.

How long should the data be stored? Who should be credited as an author?

7. What's in It for the Community?

Medical and dental care by physicians and dentists on the research team should be provided (if possible) to any interested persons from the community. However, participation in the study should not be a requirement for getting these services. All members of the community should have access to the physicians and dentists without feeling any coercion to participate in the study. In most of the LBA projects, a physician or several physicians accompanied the research team and provided health care to any community member who wanted and needed their services. There may be accreditation problems in some countries that prevented a KU team physician or dentist from practicing their craft. Under those circumstances, local, certified specialists are employed to provide the necessary services. Although our dentist, Dr. Ted Rebich, accompanied the research team to Ireland, he was not legally permitted to practice dentistry on local, itinerant populations. In this case, a local, certified dentist (Dr. Liam Convery) treated the Traveler community's dental needs.

In the Mennonite research program on biological aging, three physicians collected health histories and drew blood samples from participants. One of the physicians, Dr. Peter Hiebert, was a resident of the Goessel community while his physician son Dr. David Hiebert, a radiologist from Lawrence, Kansas, was part of the research team. The presence of a research team of specialists, exercise physiologists, biological and cultural anthropologists, and physicians generated community interest in biological aging, and more than 1,200 Mennonites from three congregations initially participated in the study. Since metabolic blood chemistry panels were assessed for each participant, the results of these tests were sent to the primary care physicians as specified by the participants.

Five graduate students from the University of Kansas received field experiences with Mennonite populations and wrote doctoral dissertations on various aspects of the aging process. Three of the dissertations focused on biological aging and two dissertations described the social organization and culture of the Mennonite communities. Several M.A. theses were based on the fieldwork.

In 2004, a follow-up study of the Mennonite biological aging and nutrition was made possible by the availability of funds from the Attorney General's Office. These funds were the result of a lawsuit filed by the Attorney General's Office against a number of food supplement companies for misrepresenting the efficacy of their products and ultimately made available for research on Kansas communities involving health and nutrition. Fortunately, the Laboratory of Biological Anthropology had in residence two key researchers with expertise in nutrition and molecular genetics (Dr. M. J. Mosher from KU and Dr. Dario Demarche from Cordoba, Argentina). M. J. Mosher completed her Ph.D. dissertation at KU on the Buriyat populations of Siberia with an emphasis on nutrition and genetics. She wrote a major portion of the successful grant application, "Lipids, Health and Coronary Heart Disease," to the Attorney General's Settlement Fund and received $132,120 for a two-year study. She also organized a research team of graduate students from the University of Kansas to update research completed two decades earlier (Mosher et al., 2016). While on a postdoctoral fellowship at the Laboratory of Biological Anthropology, Dr. Demarchi completed research on the relationship of apolipoprotein variation, lipid levels, and survivorship among the Kansas Mennonites (Demarchi et al., 2005).

Camelot Affair

A social science research project (Camelot) was funded in 1964 by the US Army to assess the causes of conflict between national groups. Chile was to be the test case, but the project was cancelled in 1965 after congressional hearings. Professor Hugo Nutini (University of Pittsburgh) was a key consultant to this project. He asked for permission to approach Chilean social scientists who exposed this Army anti-insurgency project to the Chilean legislature. Following this exposure, Hugo Nutini became a persona non-gratia in Chile (despite the fact that he had won a bronze medal for Chile in the Olympic Games), and for a time all anthropological research was forbidden.

Potential Dangers and Risks

In some cases, field research in a particular community is not possible because of ongoing pandemics, political unrest, extreme hostility, or lack of interest by the community. Currently travel restrictions due to COVID-19 persist in a number of countries. Vaccinations, the wearing of facemasks, and keeping distance from research subjects are necessary to lower risk of infection. Before the pandemic started in Peru, the LBA in collaboration with a Peruvian Medical School and the Diabetes Center at KU began a research program on Type 2 diabetes mellitus. A research team consisting of social anthropologist Professor Bart Dean, physician and director of the Diabetes Center at KUMC David Robbins, and KU graduate student and biostatistician Chad Gerhold started a pilot study in a settlement located in lowland Amazonia. I was unable to accompany the research team because of ill health. Blood samples were drawn by venipuncture by Peruvian nurses, and blood glucose levels were characterized by a local medical center. Individual diabetics and their families were identified, and DNA was extracted. However, because of the COVID-19 pandemic, shipping the DNA samples to US laboratories for sequencing was not possible at that time. The samples were stored in freezers at an American laboratory in Lima but recently shipped to the United States. In addition to potential risks to participants in the study, in a pandemic situation, risk to members of the research team must also be considered.

Social and Political Instability

Some of the Old Order Mennonites were relocated from Canada to Cuauhtemoc, in northern Mexico. However, currently it is extremely dangerous to conduct field research in that region of Mexico due to the violent drug wars among the cartels, gangs, and the Mexican government. Alternative solution: collaborating with individual Old Order Mennonites who have relocated to a safer location. While it was dangerous to work with Mexican Mennonites, it was possible to collaborate with Mennonites who migrated to Garden City in southwestern Kansas to escape ongoing violence from narcotic traffickers along the US-Mexican border.

Unfortunately, it is no longer possible to resample the baboon colonies of Sukhumi, Abkhasia, described in Chapter 8. Warfare between Abkhasia and Georgia over political control resulted in the destruction of the Sukhumi Primate Center and the deaths of most of the baboon colony. A few nonhuman primate species were smuggled by the research staff out of Abkhasia across the border to southern Russia.

Conclusion

Is any field research appropriate and ethical? Do researchers take advantage of individuals who assume that there is a direct benefit to themselves? Some researchers pay the research subjects for the time spent on the project. However, this practice has been criticized as preventing potential subjects from rejecting participation because they are economically strapped and cannot afford to pass up the opportunity to earn some money.

Ethical field research of the future must include an equal partnership between the indigenous communities and the scientists. Such collaborations will provide local scientists with the necessary field team experience and prepare them for initiating their own research programs. The cultures of science and that of the indigenous people must include communication, consideration, and sensitivity to each other.

APPENDIX A

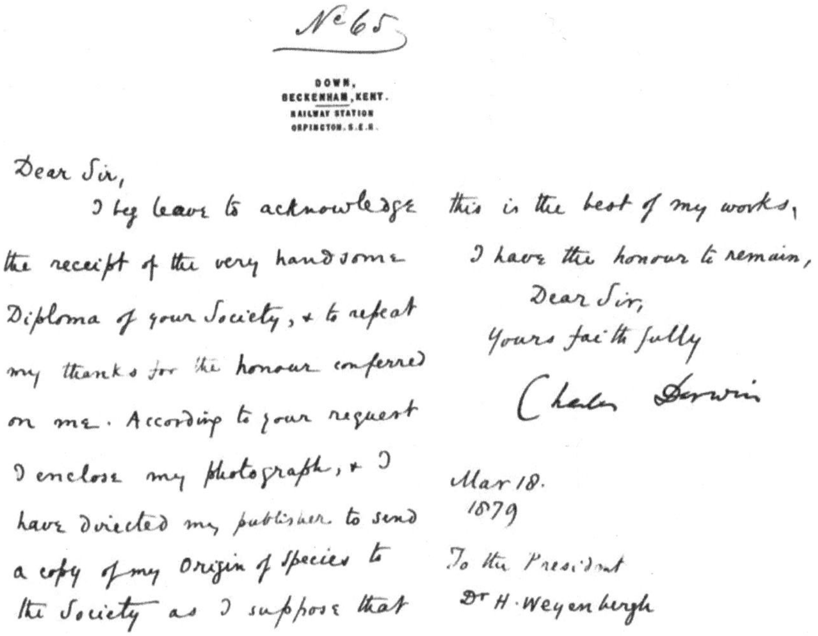

No 65

DOWN,
BECKENHAM, KENT.
RAILWAY STATION
ORPINGTON. S.E.R.

Dear Sir,

I beg leave to acknowledge the receipt of the very handsome Diploma of your Society, & to repeat my thanks for the honour conferred on me. According to your request I enclose my photograph, & I have directed my publisher to send a copy of my Origin of Species to the Society as I suppose that this is the best of my works,

I have the honour to remain,

Dear Sir,

Yours faithfully

Charles Darwin

Mar 18.
1879

To the President
Dr H. Weyenbergh

Figure A.1 Letter from Charles Darwin.

APPENDIX B

List of Former Graduate Students

Michael H. Crawford has chaired 42 dissertation committees and a total of 31 M.A. thesis committees and 23 postdoctoral fellows at the University of Kansas. The following graduate students obtained their Ph.D.s under his supervision. A listing of their present academic appointments is included:

1. Linda Klepinger, 1972, University of Illinois, Professor Emerita
2. Robert A. Halberstein, 1973, University of Miami, Former Chair, Professor, Emeritus
3. Paul Lin*, 1973, University of Chicago, Research Director of Psychiatric Institute. Retired.
4. Paul Sciulli, 1974, Ohio State University (he completed Ph.D. at Pittsburgh after Dr. Crawford moved to Kansas) Emeritus Professor
5. Douglas Ubelaker (co-chaired Ph.D. committee with T. D. Stewart), 1973, Smithsonian Institution, Chair of Division
6. Stewart Shermis*, 1974, Long Beach State University, died
7. Francis Lees, 1975, SUNY-Albany, Rockefeller University, Chief of Communications at the Museum of Natural History, NY, Retired, Professor Emeritus
8. Martin Nickels, 1975, Illinois State University, Professor Emeritus
9. Kenneth Turner*, 1976, University of Alabama, Associate Professor, Director of a museum in Oklahoma
10. Robert Baume, 1981, Connecticut Health Department, Chief Researcher, retired
11. Dennis O'Rourke, 1980, Foundation Professor of Anthropology, University of Kansas
12. Pamela J. Byard, 1981, Case Western Reserve University Medical School, Department of Pediatrics, Associate Professor, currently in the private sector
13. Janis Hutchinson, 1984, University of Houston, Professor, Former President of the Black Anthropologists Association, Emerita
14. Laurine Oberdieck Rogers, 1984, Idaho State University, Iowa State University, Adjunct Associate Professor
15. Lorena Madrigal, 1988, Southern Florida University, Professor, former President of the American Association of Physical Anthropology
16. Sueb-sak Sirijaraya*, 1988, University in Bangkok, died recently
17. Meredith Uttley, 1991, Idaho State, Lander University, Professor and Chair
18. Anthony G. Comuzzie, 1993, Southwest Foundation for Biomedical Research, Scientist (equivalent to Professor), currently President of Obesity Society
19. Ravi Duggirala, 1995, University of Texas at San Antonio Medical Center, Associate Professor of Medicine and Genetic Epidemiology, currently Professor at University of Texas Medical School, SW Foundation for Biomedical

Research—former President of the American Association of Anthropological Genetics

20. Rector Arya, 1999, University of Texas at San Antonio Medical Center, currently Assistant Professor in Genetic Epidemiology
21. Lisa Martin, 1999, Southwest Foundation for Biomedical Research, San Antonio, postdoctoral fellow; currently Professor in Genetic Epidemiology, University of Cincinnati
22. Joseph McComb, 1999, Merck Pharmaceutical Company, NJ—scientist
23. Kari North, 2000, Southwest Foundation for Biomedical Research, San Antonio, postdoctoral fellow; currently Professor, University of North Carolina, Chapel Hill, Department of Epidemiology
24. Sobha Puppala, 2001, Case Western Reserve University, postdoctoral fellow, scientist, SW Foundation for Biomedical Research, currently Assistant Professor, Wake Forest University
25. M. J. Mosher, 2002 postdoctoral fellow, Laboratory of Biological Anthropology, University of Kansas 2002–2003; postdoctoral fellow, Department of Epidemiology, University of North Carolina, Chapel Hill, Associate Professor, Western Washington University, Emerita
26. Rohina Rubicz, 2007 postdoctoral fellow, scientist, Department of Genetics, Southwest Foundation for Biomedical Research, San Antonio; currently scientist at Hutchinson Center, Seattle
27. Phillip E. Melton, 2008 postdoctoral fellow (Cowles Fellowship), Department of Genetics, Southwest Foundation for Biomedical Research, San Antonio. Associate Professor, Western Australian University, Perth, Australia
28. Mark Zlojutro, 2008, Department of Genetics, Southwest Foundation for Biomedical Research, San Antonio, University of Texas, Assistant Professor
29. Kristin L. Young, 2009, postdoctoral fellow, University of Kansas Medical Center, Kansas City, KS, postdoctoral fellow, University of North Carolina—Chapel Hill, currently Research Assistant Professor
30. Jay Sarthy, MD, Self-Fellow, July 16, 2009, defended dissertation in Genetics Program. Northwestern Medical School, Fall 2009, Ph.D.; currently faculty at University of Washington.
31. Geetha Chittoor, 2009, Genetic Epidemiology of Blood Pressure in Mexican Americans of San Antonio, postdoctoral fellowship (Cowles Fellowship) at SFBR, San Antonio. University of North Carolina; currently Geisinger Institute, PA
32. Anne Justice, 2011, Genetic structure of Ch'orti Maya from eastern Guatemala, postdoctoral fellow at University of North Carolina; currently Geisinger Institute, PA
33. Christine Phillips Krawczak—Genetics Program, 2012, postdoctoral fellow, Mayo Clinic, Rochester, Minnesota
34. Norberto Baldi Salas, 2013, Assistant Professor, University of Costa Rica
35. Jasem Theyab, 2013, Assistant Professor, University of Kuwait
36. Kristine G. Beaty, 2017 postdoctoral fellow, University of Oklahoma, Laboratory Director, KU
37. Delisa Phillips, 2017, private sector
38. Melody Ratcliffe, 2018, Tennessee
39. Randy David, 2019, San Francisco

References

Adalid-Sainz, C., R. Barquera, M. H. Crawford, A. Lona-Sanchez, S. Clayton, E. Arrieta-Bolanos, H. Delgado-Aguirre, L. Gonzalez-Medina, H. Pacheco-Ubaldo, D. I. Hernandez-Zaragoza, A. Bravo-Acevedo, N. Escareno-Montiel, J. Moran-Martinez, M. D. R. Gonzalez-Martinez, Y. Jaramillo-Rodriguez, A. Salgado-Adame, F. Juarez-de la Cruz, J. Zuniga, C. Beuter-Mendez, and J. Granados. 2019. Genetic Diversity of HLA System in Three Populations from Coahuila, Mexico: Torreon, Saltillo and Rural Coahuila, Mexico. *Human Immunology* 81(19): 492–495.

Aguirre Beltran, G. 1944. The Slave Trade in Mexico. *The Hispanic American Historical Review* 24: 412–431.

Aguirre, J. D. 1976. Tlaxcaltecan Colonization and Its Influence in Northern Mexico. In *The Tlaxcaltecans: Prehistory, Demography, Morphology and Genetics*, ed. Michael H. Crawford, vol. 7, pp. 35–38. Lawrence: University of Kansas Publications in Anthropology.

Alexander, D. H., J. Novembre, and K. Lange. 2009. Fast Model-Based Estimation of Ancestry in Unrelated Individuals. *Genome Research* 19: 1655–1664.

American Anthropological Association Ethics Statement. 2012. Principles of Professional Responsibility. Washington, DC: Author.

Ammerman, A. J., and L. L. Cavalli-Sforza. 1984. *The Neolithic Transition and the Genetics of the Populations in Europe.* Princeton, NJ: Princeton University Press.

Arnaiz-Villena, A., E. Gomez-Casado, and J. Martinez-Laso. 2002. Population Genetic Relationships between Mediterranean Populations Determined by HLA Allele Distribution and Historic Perspective. *Tissue Antigens* 60(2): 111–121.

Baldi Salas, N. 2013. *Genetic Structure and Biodemography of the Rama Amerindians from the Southern Caribbean Coast of Nicaragua.* Ph.D. dissertation, University of Kansas.

Ballard, J. G. 1985. *Empire of the Sun.* London: Panther Books.

Bamshad, M., A. E. Fraley, M. H. Crawford, R. L. Cann, B. R. Busi, J. H. Naidu, and L. Jorde. 1996. MtDNA Variation in Caste Populations of Andhra Pradesh, India. *Human Biology* 68(1): 1–20.

Baume, R. M., and M. H. Crawford. 1978. Discrete Dental Traits in Four Tlaxcaltecan Mexican Populations. *American Journal of Physical Anthropology* 49: 351–360.

Baume, R. M., and M. H. Crawford. 1980. Discrete Dental Trait Asymmetry in Mexican and Belizean Groups. *American Journal of Physical Anthropology* 52: 315–322.

Beaty, K. G. 2017. *Forced Migration and Population Expansion. The Genetic Story of the Garifuna.* Ph.D. dissertation, University of Kansas.

Beaty, K. G., M. J. Mosher, M. H. Crawford, and P. Melton. 2016. Paternal Genetic Structure in Contemporary Mennonite Communities from the American Midwest. *Human Biology* 88(2): 95–108.

Belmont Report. 1978. Report of the National Commission for the Protection of Human Subjects of Biomedical and Behavioral Research. US Department of Health, Education and Welfare. *Federal Register* 44(76): 23191–23197.

Belmont Report. April 18, 1979. Department of Health, Education, and Welfare. The National Commission for the Protection of Human Subjects and Biomedical and Behavioral Research.

Benet, S. 1976. *How to Live To Be 100: The Life-Style of the People of the Caucasus.* New York: Dial Press.

Benn-Torres, J., A. Stone, and R. Kittles. 2013. An Anthropological Genetic Perspective on Creolization in the Anglophone Caribbean. *American Journal of Physical Anthropology* 15: 135–143.

Benn-Torres, J., M. G. Vilar, G. A. Torres, et al. 2015. Genetic Diversity in the Lesser Antilles and Its Implications for the Settlement of the Caribbean Basin. *PLoS* 10(10): e0139192.

Bernstein, F. 1931. Verteilung der Blutgruppen und thre anthropologische Bedeutung. In *Comitato Italiano per Studio dei Problemi della Populazionne*, pp. 227–243. Rome: Instituto Polygrafico dello Stato Roma.

Bertorelle, G., and G. Barbujani. 1995. Analysis of DNA Diversity by Spatial Autocorrelation. *Genetics* 140: 811–819.

Black, L. T. 1981. Volcanism as a Factor in Human Ecology: The Aleutian Case. *Ethnohistory* 28(4): 313–333.

Blattner, W. A., V. S. Kalyanaraman, M. Robert-Guroff, T. Andrew Lister, D. A. G. Galton, P. S. Sarin, M. H. Crawford, D. Catovsky, M. Greaves, and R. C. Gallo. 1982. The Human Type-C Retrovirus, HTLV, in Blacks from the Caribbean Region and Relationship to Adult T-Cell Leukemia/Lymphoma. *International Journal of Cancer* 30: 257–264.

Bohn Gmelch, S., and G. Gmelch. 1976. The Emergence of an Ethnic Group: The Irish Tinkers. *Anthropological Quarterly* 49(4): 225–238.

Boyd, J. T. 2014. *Peopling of the Americas: ABO Blood Group Haplotypes as an Indicator of Native American Origins and Migration from Siberia.* M.A. thesis, University of Kansas.

Brennan, E. 1983. Factors Underlying Decreasing Fertility among the Garifuna of Honduras. *American Journal of Physical Anthropology* 60: 177.

Byard, P. J., and M. H. Crawford. 1991. Founder Effect and Genetic Diversity on St. Lawrence Island, Alaska. *Homo* 41: 219–227.

Byard, P. J., and F. C. Lees. 1981. Estimating the Number of Loci Determining Skin Color in a Hybrid Population. *Annals of Human Biology* 8: 49–58.

Byard, P. J., M. S. Schanfield, and M. H. Crawford. 1983. Admixture and Heterozygosity in West Alaskan Populations. *Journal of Biosocial Science* 15(2): 207–216.

Calafell, F., and J. Bertranpetit. 1994. Mountains and Genes: Population History of the Pyrenees. *Human Biology* 66(5): 823–842.

Cavalli-Sforza, L. L., and A. W. F. Edwards. 1967. Phylogenetic Analysis: Models and Estimation Procedures. *American Journal of Human Genetics.* 19: 233–257.

Clavell, J. 1975. *Shogun: A Novel of Japan.* New York: Macmillan.

Cockburn, T. A. 1963. *The Evolution and Eradication of Infectious Disease.* Baltimore: Johns Hopkins Press.

Comuzzie, A. G. 1993. *Genomic, Genetic and Morphological Variation in a Sample of Modern Evenki and Their Relationship with Other Indigenous Siberian Populations.* Ph.D. dissertation, University of Kansas.

Comuzzie, A. G., and M. H. Crawford. 1990. Biochemical Heterozygosity and Morphological Variability: Interpopulational versus Intrapopulational Analyses. *Human Biology* 62: 101–112.

Crawford, M. H. 1966. Hemoglobin Polymorphism in *Macaca nemestrina. Science* 154: 398–399.

Crawford, M. H. 1967. *A Re-examination of the Taxonomy and Phylogeny of the Hominoidea Based upon Experimental Data.* Ph.D. dissertation, University of Washington.

Crawford, M. H. 1975. Genetic Affinities and Origin of the Irish Tinkers. In *Biosocial Interrelations in Population Adaptations*, edited by E. Watts, F. Johnston, and G. W. Lasker, pp. 93–103. The Hague: Mouton Press.

Crawford, M. H. 1976. Introduction: Problems and Hypotheses. In *The Tlaxcaltecans: Prehistory, Demography, Morphology and Genetics*, edited by M. H. Crawford, vol. 7, pp. 1–5. Lawrence: University of Kansas Publications in Anthropology.

Crawford, M. H. 1980. The Breakdown of Reproductive Isolation in an Alpine Genetic Isolate: Acceglio, Italy. In *Population Structure and Genetic Disorders*, edited by A. Eriksson, H. R. Forsius, H. R. Nevanlinna, P. L. Workman, and R. J. Norio, pp. 57–71. New York: Academic Press.

Crawford, M. H. 1983. The Anthropological Genetics of the Black Caribs (Garifuna) of Central America. *Yearbook of Physical Anthropology* 26: 161–192.

Crawford, M. H., ed. 1984. *Current Developments in Anthropological Genetics. Vol. 3. Black Caribs. A Case Study in Biocultural Adaptation.* New York: Plenum Press.

Crawford, M. H. 1998. *Origins of Native Americans: Evidence from Anthropological Genetics*. Cambridge University Press.

Crawford, M. H., ed. 2000. *Different Seasons: Biological Aging among the Mennonites of the Midwestern United States.* University of Kansas Publications in Anthropology, vol. 21. Lawrence: University of Kansas Press.

Crawford, M. H. 2005. Genetics of Biological Aging in Mennonites of Midwestern United States. *Przeglad Antropologiczny Anthropological Review* 68: 3–18.

Crawford, M. H. 2006. Who Are We? Aleut Research Program (1999–2006). *The Aleutian Current. Aleut Corporation Newsletter* 35(4): 6–7.

Crawford, M. H., ed. 2007a. *Anthropological Genetics: Theory, Methods and Applications.* New York: Cambridge University Press.

Crawford, M. H. 2007b. Foundations of Anthropological Genetics. In *Anthropological Genetics: Theory, Methods and Applications*, edited by M. H. Crawford, pp. 1–16. New York: Cambridge University Press.

Crawford, M. H. 2007c. Genetic Structure of Circumpolar Populations: A Synthesis. *American Journal of Human Biology* 19: 203–217.

Crawford, M. H. 2007d. The Importance of Field Research in Anthropological Genetics: Methods, Experiences and Results. In *Anthropological Genetics. Theories, Methods and Applications*, pp. 79–111. New York: Cambridge University Press.

Crawford, M. H. 2010a. Origin of Aleuts and the Genetic Structure of Populations of the Archipelago: Molecular and Archaeological Perspectives. *Human Biology* 82(5–6): 695–717.

Crawford, M. H. 2015. Genetic Structure and Its Implications for Genetic Epidemiology: Aleutian Island Populations. In *Genomics in Human and Non-Human Primates*, edited by R. Duggirala, A. G. Comuzzie, S. Williams-Blangero, and C. Cole, pp. 129–140. New York: Springer Life Sciences.

Crawford, M. H. 2016. Introduction to the Fortieth Anniversary of the Founding of the Laboratory of Biological Anthropology. Special Issue. *Human Biology* 88(2): 93–94.

Crawford, M. H. 2018. Raymond Pearl. *The International Encyclopedia of Biological Anthropology.* Edited by Wenda Trevathan. Hoboken, NJ: Wiley Blackwell.

Crawford, M. H., S. Alden, R. E. David, and K. Beaty. 2021. Unangan (Aleut) Migrations: Causes and Consequences. In *Human Migration: Biocultural Perspectives*, edited by L. Munoz-Moreno and M. H. Crawford, pp. 20–31. New York: Oxford University Press.

Crawford, M. H., and E. J. Devor. 1980. Population Structure and Admixture in Tlaxcaltecan Populations. *American Journal of Physical Anthropology* 52: 485–490.

Crawford, M. H., D. D. Dykes, K. Skradski, and H. F. Polesky. 1979. Gene Flow and Genetic Microdifferentiation of a Transplanted Tlaxcaltecan Indian Population: Saltillo. *American Journal of Physical Anthropology* 50(3): 401–413.

Crawford, M. H., D. Dykes, J. Mielke, and H. F. Polesky. 1981. Population Structure of Circumpolar Alaskan and Siberian Indigenous Communities. *American Journal of Physical Anthropology* 55: 167–186.

Crawford, M. H., D. D. Dykes, K. Skradsky, and H. Polesky. 1984. Blood Group, Serum Protein, and Red Cell Enzyme Polymorphisms, and Admixture among the Black Caribs and Creoles of Central America and the Caribbean. In *Current Developments in Anthropological Genetics: Black Caribs: A Case Study in Biocultural Adaptation*, edited by M. H. Crawford, vol. 3, pp. 303–333. New York: Plenum Press.

Crawford, M. H., and G. Gmelch. 1974. The Human Biology of the Irish Tinkers: Demography, Ethnohistory and Genetics. *Social Biology* 21: 321–331.

Crawford, M. H., N. L. Gonzalez, M. S. Schanfield, et al. 1981. The Black Caribs (Garifuna) of Livingston, Guatemala: Genetic Markers and Admixture Estimates. *Human Biology* 53(1): 87–103.

Crawford, M. H., T. Koertvelyessy, M. Pap, K. Szilagyi, and R. Duggirala. 1999. The Effects of a New Political Border on the Migration Patterns and Predicted Kinship (PHI) in a Subdivided Hungarian Agricultural Population: Tiszahat. *Homo* 50(3): 201–210.

Crawford, M. H., and G. W. Lasker, eds. 1989. Foundations of Anthropological Genetics. *Human Biology* 61(5–6): v–vi.

Crawford, M. H., R. Lisker, and R. Perez Briceno. 1976. Genetic Microdifferentiation of Two Transplanted Tlaxcaltecan Populations. In *The Tlaxcaltecans: Prehistory, Demography, Morphology and Genetics*, edited by M. H. Crawford, vol. 7, pp. 169–175. Lawrence: University of Kansas Publications in Anthropology.

Crawford, M. H., and J. H. Mielke, eds. 1982. *Current Developments in Anthropological Genetics. Vol. II. Ecology and Population Structure*. New York: Plenum Press.

Crawford, M. H., and D. H. O'Rourke. 1978. Inbreeding, Lymphoma, Genetics and Morphology of the Papio hamadryas Colony of Sukhumi. *Journal of Medical Primatology* 7(6): 355–360.

Crawford, M. H., D. H. O`Rourke, D. D. Dykes, L. A. Yakovleva, A. F. Voevodin, B. Lapin. and H. F. Polesky. 1984. Inbreeding, Heterozygosity and Lymphoma Risk among the Baboons (*Papio hamadryas*) of the Sukhumi Primate Center. *American Journal of Primatology* 63: 143–153.

Crawford, M. H., C. Phillips-Krawczak, C. G. Beaty, and N. Boaz. 2021. Migration of Garifuna: Evolutionary Success Story. In *Human Migration: Biocultural Perspectives*,edited by L. Munoz-Moreno and M. H. Crawford, pp. 153–168. New York: Oxford University Press.

Crawford, M. H., and L. Rogers. 1982. Population Genetic Models in the Study of Aging and Longevity in a Mennonite Community. *Social Science and Medicine* 18: 149–153.

Crawford, M. H., R. Rubicz, and M. Zlojutro. 2010. Origins of the Aleuts and the Genetic Structure of Populations of the Archipelago: Molecular and Archaeological Perspective. *Human Biology* 82(5–6): 695–718.

Crawford, M. H., and D. West. 2012. Evolutionary Consequences of Human Migration: Genetic, Historic and Archaeological Perspectives in the Caribbean and Aleutian Islands. In *Causes and Consequences of Human Migration*, pp. 65–86. New York: Cambridge University Press.

Crawford, M. H., D. West, and D. H. O'Rourke, eds. 2010. Double-Special-Issue on: The Aleuts: Origins, Culture and Genetics. *Human Biology* 82(5–6): 481–764.

Crawford, M. H., J. T. Williams, and R. Duggirala. 1997. Genetic Structure of Indigenous Populations of Siberia. *American Journal of Physical Anthropology* 104(2): 177–192.

Crawford, M. H., P. L. Workman, C. McLean, and F. C. Lees. 1976. Admixture Estimates and Selection in Tlaxcala. In *The Tlaxcaltecans: Prehistory, Demography, Morphology and Genetics*, vol. 7, pp. 161–168. Lawrence: University of Kansas Publications in Anthropology.

Cummins, H. 1930. Dermatoglyphics in Indians of Southern Mexico and Central America. *American Journal of Physical Anthropology* 15: 123–130.

Cummins, H., and D. Midlo. 1943. *Finger Prints, Palms and Soles*. Philadelphia: Blackiston.

Custodio, R., and R. Huntsman. 1984. Abnormal Hemoglobins among the Black Caribs. In *Current Developments in Anthropological Genetics: Black Caribs, a Case Study in Biocultural Adaptation*, edited by M. H. Crawford, vol. 3, pp. 335–343. New York: Plenum Press.

Dahlberg, A. A. 1956. *Materials for the Establishment of Standards for the Classification of Tooth Characters, Attributes and Techniques in Morphological Studies of Dentition*. Zoller Laboratory of Dental Anthropology. Chicago: University of Chicago.

Davidson, W. V. 1984. The Garifuna in Central America: Ethnohistorical and Geographical Foundations. In *Current Developments in Anthropological Genetics: Black Caribs. A Case Study in Biocultural Adaptation*, edited by M. H. Crawford, vol. 3, pp. 13–33. New York: Plenum Press.

Davila Aguirre, J. J. 1976. Tlaxcaltecan Colonialization and Its Influence in Northern Mexico. In *The Tlaxcaltecans: Prehistory, Demography, Morphology and Genetics*, edited by M. H. Crawford, , vol. 7, pp. 35–37. Lawrence: University of Kansas Publications in Anthropology.

Davis, D. D., and C. Goodwin. 1990. Island Carib Origins: Evidence and Nonevidence. *American Antiquity* 55(1): 37–48.

Davis, R. S., and R. A. Knecht. 2010. Continuity and Change in the Eastern Aleutian Archaeological Sequence. *Human Biology* 82(5–6): 507–524.

Demarchi, D., M. J. Mosher, and M. H. Crawford. 2005. Apolipoproteins (Apoproteins) and LPL Variation in Mennonite Populations of Kansas and Nebraska. *American Journal of Human Biology* 17(5): 593–600.

Derbeneva, O. A., R. I. Sukernik, N. V. Voldodko, S. H. Hosseini, and M. Y. Lott. 2002. Analysis of Mitochondrial DNA Diversity in the Aleuts of the Commander Islands and Its Implications for the Genetic History of Beringia. *American Journal of Human Genetics* 71(2): 415–421.

Devor, E. J. 2000. Age and the Quantitative Genetics of Neuromuscular Performance. In *Different Seasons: Biological Aging among Mennonites of Midwestern United States*, pp. 69–76. Publications in Anthropology, vol. 21. Lawrence: University of Kansas.

Devor, E. J., M. H. Crawford, and W. Osness. 1985. Neuromuscular Performance in a Kansas Mennonite Community: Age and Sex Effects in Performance. *Human Biology* 57(2): 197–212.

Devor, E. J., M. McGue, M. H. Crawford, and P. M. Lin. 1986a. Transmissible and Non-Transmissible Components of Anthropometric Variation in the Alexanderwohl Mennonites. I. Description and Familial Correlations. *American Journal of Physical Anthropology* 69(1): 71–82.

Devor, E. J., M. McGue, M. H. Crawford, and P. M. Lin. 1986b. Transmissible and Non-Transmissible Components of Anthropometric Variation in the Alexanderwohl Mennonites. II. Resolution by Path Analysis. *American Journal of Physical Anthropology* 69(1): 83–92.

Duggirala, R. 1995. Cultural and Genetic Determinants of Lipids and Lipoproteins in the Mennonite Community. Ph.D. dissertation, University of Kansas.

Duggirala, R., M. H. Crawford, and T. Koertvelyessy. 1992. Tarpa, Northeastern Hungary: 1780–1979. Based on Isonymy and Repeated Pairs (RP) Methods. *Journal of Indian Anthropological Society* 26: 63–75.

Duggirala, R., M. Uttley, K. K. Williams, R. Arya, J. Blangero, and M. H. Crawford. 2002. Genetic Determination of Biological Aging in the Mennonites of the Midwestern United States. *Genetic Epidemiology* 23: 97–109.

Dyck, C. J. 1993. The General Conference Mennonite Church. In *An Introduction to Mennonite History*, pp. 252–276. Scottdale, PA: Herald Press.

Estrada-Mena, B., F. J. Estrada, R. Ulloa-Arvizu, et al. 2010. Blood Group O Alleles in Native Americas: Implications in the Peopling of the Americas. *American Journal of Physical Anthropology* 142(1): 85–94.

Field, A. 1977. *Nabokov: His Life in Part.* New York: Penguin Books.

Firschein, L. I. 1961. Population Dynamics of the Sickle-Cell Trait in the Black Caribs of British Honduras, Central America. *American Journal of Human Genetics* 13(2): 233–254.

Firschein, L. I. 1984. Demographic Patterns of the Garifuna (Black Caribs) of Belize. In *Current Developments in Anthropological Genetics. Vol. 3. Black Caribs. A Case Study in Biocultural Adaptation*, pp. 67–94. New York: Plenum Press.

Fitzpatrick, S. M. 2015. The Pre-Columbian Caribbean Colonization, Population Dispersal, and Island Adaptations. *Research Gate*, November, 305–331.

Flegel, W. A. 2011. Molecular Genetics and Clinical Applications for RH. *Transfusion* 44(1): 81–91.

Galton, F. 1892. *Finger Prints.* London: MacMillan.

Gamboa, I. 1976. Malnutrition and Disease among the Tlaxcaltecans: San Pablo del Monte and Cuanalan. In *The Tlaxcaltecans: Prehistory, Demography, Morphology and Genetics*, vol. 7, pp. 145–149. Lawrence: University of Kansas Publications in Anthropology.

Gonzalez, N. L. 1969. *Black Carib Household Structure.* Seattle: University of Washington.

Gonzalez, N. L. 1984. Garifuna (Black Carib) Social Organization. In *Current Developments in Anthropological Genetics. Vol. 3. Black Caribs: A Case Study in Biocultural Adaptation*, edited by M. H. Crawford, pp. 51–65. New York: Plenum Press.

Graf, O., M. Zlojutro, R. Rubicz, and M. H. Crawford. 2010. Surname Distributions and Y-Chromosome Markers in the Aleutian Islands. *Human Biology* 82(5–6): 745–758.

Halberstein, R. A. 1973. *Evolutionary Implications of Demographic Structure of a Transplanted Population in Central Mexico.* Unpublished Ph.D. dissertation, University of Kansas.

Halberstein, R. A., M. H. Crawford, and H. G. Nutini. 1973. Historical Demographic Analysis of Indian Populations in Tlaxcala, Mexico. *Social Biology* 20: 40–50.

Harpending, H., and R. Ward. 1982. Chemical Systematics and Human Populations. In *Biochemical Aspects of Evolutionary Biology*, edited by M. H. Nitecki, pp. 213–256. Chicago: University of Chicago Press.

Harvey, R. G., M. J. Godber, A. C. Kopec, A. R. Mourant, and D. Tills. 1969. Frequency of Genetic Traits in the Caribs of Dominica. *Human Biology* 41(3): 342–364.

Henry, E. R. 1900. *Classification and Uses of Finger-Prints*. London: Routledge.

Herrera-Paz, E. F., M. Mattamoros, and A. Carracedo. 2010. The Garifuna (Black Carib) People of the Atlantic Coasts of Honduras. *American Journal of Human Biology* 22: 36–44.

Holt, S. B. 1968. *The Genetics of Dermal Ridges*. Springfield, IL: Charles C. Thomas.

Hutchinson, J., and M. H. Crawford. 1981. Genetic Determinants of Blood Pressure Level among the Black Caribs of St. Vincent. *Human Biology* 533: 453–466.

Hutchinson, J., P. Lin, and M. H. Crawford. 1983. Factors Influencing Blood Pressure Level among the Black Caribs of St. Vincent Island. In *Current Developments in Anthropological Genetics, vol. 3. The Black Caribs*, edited by M. H. Crawford, pp. 215–237. New York: Plenum Press.

Jochelson, V. I. 1933. *History, Ethnology, and Anthropology of the Aleut*. Publication 432. Washington, DC: Carnegie Institution of Washington.

Justice, A., R. Rubicz, G. Chitoor, R. L. Jantz, and M. H. Crawford. 2010. Anthropometric Variation among Bering Sea Natives. *Human Biology* 82(5–6): 653–676.

Karafet, T., F. L. Mendez, M Meilerman, P. A. Underhill, S. L. Zegura, and M. Hammer. 2008. New Binary Polymorphisms Reshape and Increase Resolution of the Human Y Chromosomal Haplogroup Tree. *Genome Res* 18(5): 30–38.

Keith, M. H., M. V. Flinn, H. J. Durbin, T. N. Rowan, G. E. Blomquist, K. H. Taylor, J. F. Taylor, and J. E. Decker. 2021. Genetic Ancestry, Admixture, and Population Structure in Rural Dominica. *PLoS* 16(11): e0258735.

Koertvelyessy, T., M. H. Crawford, and J. Hutchinson. 1982. PTC Threshold Distributions and Age in Mennonite Populations. *Human Biology* 54(3): 636–646.

Koertvelyessy, T., M. H. Crawford, M. Pap, and K. Szilagyi. 1990. Surname Repetition and Isonymy in Northeastern Hungarian Marriages. *Human Biology* 62(4): 515–524.

Koertvelyessy, T., M. H. Crawford, M. Pap, and K. Szilagyi. 1992. The Influence of Religious Affiliation on Surname Repetition in Marriages in Tiszaszalka, Hungary. *Journal of Biosocial Science* 24(1): 113–121.

Koertvelyessy, T., M. H. Crawford, M. Pap, and K. Szilagyi. 1993. The Influence of Religious Affiliation on Surname Repetition (RP) in Marriages of Marokpapi, Hungary. *Antropologischen Anzieger* 51: 309–316.

Kolman, C. J., N. Sambuughin, and E. Bermingham. 1996. Mitochondrial DNA Analysis of Mongolian Populations and Implications for the Origin of New World Founders. *Genetics* 142: 1321–1334.

Kraus, B. S. 1969. *Dental Anatomy and Occlusion: A Study of the Masticatory System*. Philadelphia: Williams and Wilkins.

Lambeck, K., Y. Yokoyama, and T. Purcell. 2002. Into and out of the Last Glacial Maximum: Sea-Level Changes during Oxygen Isotope Stage 3 and 2. *Quarternary Science Review* 21: 343–360.

Lapin, B. 1975. Possible Ways of Viral Leukemia Spreads among the Hamadryas Baboons of the Sukhumi Monkey Colony. In *Comparative Leukemia Research 1973, Leukemogenesis*, edited by Y. Ito and R. Dutcher, pp. 75–84. Basel: Karger Press.

Lasker, G. W. 1985. *Surnames and Genetic Structure.* New York: Cambridge University Press.

Leaf, A. 1973. Getting Old. *Scientific American* 229: 45–52.

Lees, F. C. 1975. *A Numerical Analysis of the Anthropometrics of Tlaxcalan Populations.* Ph.D. dissertation, University of Kansas, Lawrence, KS.

Lees, F. C., and M. H. Crawford. 1976. Anthropometric Variation in Tlaxcaltecan Populations. In *The Tlaxcaltecans: Prehistory, Demography, Morphology and Genetics*, edited by Michael H. Crawford, pp. 61–80. University of Kansas Publications in Anthropology, vol. 7.

Lin, P. M. 1976. A Factor Analysis of the Dentition of Three Tlaxcaltecan Populations. In *The Tlaxcaltecans: Prehistory, Demography, Morphology and Genetics*, edited by Michael H. Crawford, vol. 7, pp. 93–119. Lawrence: University of Kansas Publications in Anthropology.

Lin, P. M. 1984. Anthropometry of Black Caribs. In *Current Developments in Anthropological Genetics. Vol. 3. Black Caribs*, edited by M. H. Crawford, pp. 189–214. Plenum Press.

Lin, P. M., V. Bach-Enciso, M. H. Crawford, J. Hutchinson, D. Sank, and B. S. Firschein. 1984. Quantitative Analyses of Dermatoglyphic Patterns of Black Carib Populations of Central America. In *Current Developments in Anthropological Genetics. Vol. 3, Black Caribs. A Case Study of Biocultural Adaptation*, edited by M. H. Crawford, pp. 241–268. New York: Plenum Press.

Longhofer, J. 1986. *Land, Household and Community: A Study of Alexanderwohl Mennonites*. Ph.D. dissertation, University of Kansas.

Lopes, F. L., L. Hou, A. B. W. Boldt, L. Kassem, V. M. Alves, A. E. Nardi, and F. J. McMahon. 2016. Finding Rare, Disease-Associated Variants in Isolated Groups: Potential Advantages of Mennonite Populations. *Human Biology* 88: 109–120.

MacMahon, B. 1971. A Portrait of Tinkers. *Natural History* 80: 24–35, 104–109.

Malecott, G. 1969. *The Mathematics of Heredity*. San Francisco: Freeman.

Margulis, L. 1975. Symbiotic Theory of the Origin of Eukaryotic Organelles: Criteria for Proof. *Symposia of the Society for Experimental Biology* 29: 21–38.

Marsh, W. L., R. S. K. Chaganti, F. H. Gardener, P. C. Nowell, and J. German. 1974. Mapping Human Autosomes: Evidence Supporting Assignment of Rhesus to the Short Arm of Chromosome no. 1. *Science* 183(4128): 966–968.

Martin, L. J. 1993. *The Genetic and Environmental Components of Thyroxine Variation in the Mennonites of Kansas and Nebraska.* M.A. thesis, University of Kansas.

Martin, L. I., K. E. North, and M. H. Crawford. 2000. The Origins of Irish Travelers and the Genetic Structure of Ireland. *Annals of Human Biology* 25: 453–465.

Mazess, R. B., and S. H. Forman. 1979. Longevity and Age Exaggeration in Vilcabamba, Ecuador. *Journal of Gerontology* 34: 94–98.

McCartney, A. P. 1984. Prehistory of the Aleutian Region. In *Handbook of North American Indians, Vol. 5: Arctic*, edited by D. Damas, 119–135. Washington, DC: Smithsonian Institution Press.

McCartney, A. P., and W. Veltre. 1996. Anagula Core and Blade Site. In *American Beginnings: The Prehistory and Paleoecology of Beringia*, edited by F. H. West, pp. 443–450. Chicago: University of Chicago Press.

McComb, J. 1996. The Effects of Unique Historical Events in the Gene Pool of the Altai-Kizhi: A Study of Five Variable Number Tandem Repeats (VNTR) Loci. M.A. thesis, University of Kansas.

Melton, P. E. 2012. Mennonite Migrations: Genetic and Demographic Consequences. In *Causes and Consequences of Human Migration: An Evolutionary Perspective*, edited by M. H. Crawford and B. C. Campbell, pp. 299–316. New York: Cambridge University Press.

Melton, P. E., M. J. Mosher, R. Rubicz, M. Zlojutro, and M. H. Crawford. 2010. Mitochondrial DNA Diversity in Mennonites from the Midwestern United States. *Human Biology* 82: 267–289.

Meltzer, D. J. 2009. *First Peoples in a New World Colonizing Ice Age*. Berkeley: University of California.

Mielke, J. H., and M. H. Crawford, eds. 1980. *Current Developments in Anthropological Genetics*, Vol. 1. Theory and Methods. New York: Plenum Press.

Monmonier, M. 1973. Maximum-Difference Barriers: An Alternative Regionalization Method. *Geographic Analysis* 3: 245–261.

Morton, N. E. 1973. Prediction of Kinship from Genealogies. In *Genetic Structure of Population*, edited by N. E. Morton, pp. 89–91. University of Hawaii Press.

Mosher, M. J. 2002. *The Genetic Architecture of Plasma Lipids in the Buryats: As Ecogenetic Approach*. Ph.D. dissertation, University of Kansas.

Mosher, M. J., P. E. Melton, P. Stapleton, M. S. Schanfield, and M. H. Crawford. 2016. Methylation across the Leptin Core Promoter in Four Diverse Asian and North American Populations. *Human Biology* 88(2): 121–135.

Munoz, L., and M. H. Crawford, eds. 2021. *Human Migration: Biocultural Perspective*. New York: Oxford University Press.

Nabokov, V. 1955 *Lolita*. Olympia, WA: Olympia Press.

Nash, G. 2002. *The Tarasov Saga: From Russia through China to Australia*. NSW, Australia: Rosenberg.

Nava, L. 1969. *Transcendencia Historia de Tlaxcala*. Mexico City: Editorial Progresso.

Neel, J. V., and F. M. Salzano. 1964. A Prospectus for the Genetic Studies of American Indian. *Cold Spring Harbor Symposium on Quantitative Biology* 20: 85–98.

North, K. E., and M. H. Crawford. 1996. Isonymy and Repeated Pairs Analysis: The Mating Structure of Acceglio, Italy, 1889–1968. *Revista di Antropologia (Roma)* 74: 93–103.

North, K. E., L. J. Martin, and M. H. Crawford. 2000. The Origins of the Irish Travelers and the Genetic Structure of Ireland. *Annals of Human Biology* 27: 453–465.

Novelletto, A. 2007. Y-chromosome Variation in Europe: Continental and Local Processes in the Formation of the Extant Gene Pool. *Annals of Human Biology* 34(2): 139–172.

Nutini, H. G. 1976. An Outline of Tlaxcaltecan Culture, History, Ethnology and Demography. In *The Tlaxcaltecans: Prehistory, Demography, Morphology and Genetics*, edited by Michael H. Crawford, vol. 7, pp. 24–34. Lawrence: University of Kansas Publications in Anthropology.

O'Rourke, D. H. 1980. *Inbreeding Effects on Morphometric Characters in the Papio Hamadryas Colony at Sukhumi*. Ph.D. Dissertation, University of Kansas.

O'Rourke, D. H., and M. H. Crawford. 1976. Odontometric Analysis of Four Tlaxcaltecan Communities. In *The Tlaxcaltecans: Prehistory, Demography, Morphology, and Genetics*, edited by M. H. Crawford, vol. 7, pp. 81–92. Lawrence: University of Kansas Publications in Anthropology.

O'Rourke, D. H., and M. H. Crawford. 1980. Odontometric Microdifferentiation of Transplanted Mexican Indian Populations: Cuanalan and Saltillo. *American Journal of Physical Anthropology* 53(3): 421–434.

O'Rourke, D. H., R. M. Baume, J. H. Mielke and M. H. Crawford. 1984. Dental Variation in Black Cari Populations. In *Current Developments in Anthropological Genetics. Vol. 3. Black Caribs*, edited by M. H. Crawford, pp. 169–186. Plenum Press.

O'Rourke, D. H., D. L. West, and M. H. Crawford. 2010. Unangan Past and Present: The Contrasts between Observed and Inferred Histories. *Human Biology* 82(5–6): 759–764.

Pakendorf, B., and M. Stoneking. 2005. Mitochondrial DNA and Human Evolution. *Annual Review Genomics Human Genetics* 6: 165–183.

Patterson, N., A. L. Price, and D. Reich. 2006. Population Structure and Eigenanalysis. *PLOS Genetics* 2(12): e190.

Pavon-Vargas, M., M. H. Crawford, R. Barquera, et al. 2019. Genetic Diversity of HLA System in Two Populations from Tlaxcala, Mexico: Tlaxcala City and Rural Tlaxcala. *Human Immunology* 81(9): 544–546.

Phillips-Krawczak, C. 2012. *Origins and Genetic Structure of the Garifuna Population of Central America.* Ph.D. dissertation, University of Kansas.

Pichler, L., C. Fuchsberger, C. Platzer, et al. 2010. Drawing the History of the Hutterite Population on the Genetic Landscape: Inference from Y-chromosome and mtDNA Genotypes. *European Journal of Human Genetics* 18: 509.

Pollin, T. L., D. J. McBride, R. Agarwala, et al. 2007. Investigations of the Y Chromosome, Male Founder Structure and YSTR Mutation Rates in the Old Order Amish. *Human Heredity* 65: 91–104.

Price, A. L., N. J. Patterson, R. M. Plenge, M. E. Weinblatt, N. A. Shadick, and D. Reich. 2006. Principal Component Analysis Corrects for Stratification in Genome-Wide Association Studies. *Nature Genetics* 38(8): 904–909.

Puppala, S. 2000. Genetic Variation in Blood Pressure in Mennonite Populations of Kansas and Nebraska. Ph.D. dissertation, University of Kansas.

Purcell, S., B. Neal, K. Todd-Brown, et al. 2007. PLINK: A Tool Set for Whole-Genome Association on Population Based Linkage. *American Journal of Human Genetics* 81(3): 559–575.

Raff, J., J. Tackney, and D. H. O'Rourke. 2010. South from Alaska: A Pilot Study of the Genetic History on the Alaska Peninsula and the Eastern Aleutians. *Human Biology* 82(5–6): 677–694.

Rahman, S., H. Hu, E. McNeely, et al. 2008. Social and Environmental Risk Factors for Hypertension in African Americans. *Florida Public Health Review* 5: 64–72.

Raj, A., M. Stephens, and J. K. Pritchard. 2014. Structure: Variational inference of population structure in large SNP data sets. *Genetics* 197(2): 573–589.

Read, K. E. 1965 *The High Valley.* New York: Charles Scribner & Sons.

Reedy-Maschner, K. 2010. Where Did All the Aleut Men Go? Aleut Male Attrition and Related Patterns in Aleutian Historical Demography and Social Organization. *Human Biology* 82(5–6): 583–612.

Reich, D., N. Patterson, D. Campbell, et al. 2012. Reconstructing Native American Population History. *Nature* 488: 370–374.

Relethford, J. H., and M. H. Crawford. 2013. Genetic Drift and Population History of the Irish Travelers. *American Journal of Physical Anthropology* 150: 184–189.

Renwick, J. H. 1971. The Rhesus Synthetic Group in Man. *Nature* 234: 475.

Roberts, D. F. 1965. Assumption and Fact in Anthropological Genetics. *Journal of the Royal Anthropological Institute* 95: 87–103.

Roberts, D. F. 1968. Genetic Effects of Population Size Reduction. *Nature* 14: 2205.

Rogers, L. 1984. *Phylogenetic Identification of a Religious Isolate and Measurement of Inbreeding*. Ph.D. dissertation, University of Kansas.

Rogers, L., and R. A. Rogers. 2000. Mennonite History with Special Reference to Alexanderwohl and Related Congregations in Kansas and Nebraska. In *Different Seasons: Biological Aging among the Mennonites of Midwestern United States*, edited by M. H. Crawford, 7–18. Lawrence: University of Kansas.

Rubicz, R. C. 2007. *Evolutionary Consequences of Recently Founded Aleut Communities in the Commander and Pribilof Islands*. Ph.D. dissertation, University of Kansas.

Rubicz, R. C., and M. H. Crawford. 2016. Molecular Genetic Evidence for the Origins of North American Populations. In *Handbook of Arctic Archaeology*, edited by O. Mason and M. Friesen, pp. 27–50. New York: Oxford University Press.

Rubicz, R. C., P. E. Melton, and M. H. Crawford. 2007. Molecular Markers in Anthropological Genetic Studies. In *Anthropological Genetics: Theory, Methods and Applications*, edited by M. H. Crawford, pp. 141–186. Cambridge University Press.

Rubicz, R. C., P. E. Melton, V. Spitsyn, G. Sun, R. Deka, and M. H. Crawford. 2010. Genetic Structure of Native Circumpolar Populations Based on Autosomal, Mitochondrial and Y-chromosome DNA Markers. *American Journal of Physical Anthropology* 143(1): 62–74.

Rubicz, R. C., K. Melvin, and M. H. Crawford. 2002. Genetic Evidence for the Phylogenetic Relationship between Na-Dene and Yeniseian Speakers. *Human Biology* 74(6): 743–760.

Rubicz, R. C., T. Schurr, P. L. Babb, and M. H. Crawford. 2003. Mt-DNA Sequences and the Origin of the Aleuts. *Human Biology* 75(6): 809–835.

Rubicz, R. C., M. Zlojutro, G. Sun, V. Spitsyn, R. Deka, K. Young, and M. H. Crawford. 2010. Genetic Architecture of a Small, Recently Aggregated Aleut Population: Bering Island. *Human Biology* 82(506): 719–736.

Ruhlen, M. 1991. *A Guide to the World's Languages*. Stanford, CA: Stanford University Press.

Ruhlen, M. 1998. The Origin of the NaDene. *Proceedings of the National Academy of Sciences USA* 95: 13994–13996.

Sagan, L. 1967. On the Origin of Mitosing Cells. *Journal of Theoretical Biology* 14(3): 225–274.

Savinetsky, A., N. K. Kisleva, and B. F. Khassanov. 2010. Paleoenvironments-Holocene Deposits from Shemya Island. In *The People at the End of the World: The Western Aleutians Project and Archaeology of Shemya Island*, edited by D. Corbett, D. West, and C. Lefevre, pp. 71–82. Anchorage: Aurora.

Schanfield, M. S. 1976. Immunoglobin Haplotypes in Tlaxcaltecan and Other Populations. In *The Tlaxcaltecans: Prehistory, Demography, Morphology and Genetics*, vol. 7, pp. 150–154. Lawrence: University of Kansas Publications in Anthropology.

Schanfield, M. S., R. Brown, and M. H. Crawford. 1984. Immunoglobulin Allotypes in the Black Caribs and Creoles of Belize and St. Vincent Island. In *Current Developments in Anthropological Genetics: Volume 3, Black Caribs-A Case Study in Biocultural Adaptation*, edited by M. H. Crawford, pp. 345–363. New York: Plenum Press.

Schatzl, H., M. Tschikobava, D. Rose, A. Voevodin, H. Nitschko, E. Sieger, U. Busch, K. von der Helm, and B. Lapin. 1993. The Sukhumi Primate Monkey Model for Viral

Lymphomogenesis: High Incidence of Lymphomas with Presence of STLV-I and EBV-Like Virus. *Leukemia* August 7, Suppl. 2: S86–S92.

Schroeder, K. B., M. Jakobbsson, M. H. Crawford et al. 2009. Haplotypic Background of a Private Allele at High Frequency in the Americas. Molecular Biology and Evolution 26(5): 995–1016.

Schroeder, W. A. 1974 .Microchromatography of Hemoglobins IV: An Improved Procedure for the Detection of Hemoglobins S and C at Birth. *Journal of Laboratory and Clinical Medicine* 86: 528–532.

Schroeder, W. A., J. Jakway, and D. Powers. 1973. Rapid Diagnosis of Sickle Cell Disease at Birth by Microcolumn Chromatography. *Journal of Laboratory and Clinical Medicine* 82: 303–308.

Scott, E. C. 2004. *Evolution and Creationism: An Introduction*. Oakland: University of California Press.

Seielstad, M., N. Yuldasheva, N. Singh, P. Underhill, P. Oefner, P. Snen, and R. S. Wells. 2003. A Novel Y-chromosome Variant Puts an Upper Limit on the Timing of the First Entry into the Americas. *American Journal of Human Genetics* 73(3): 700–705.

Smithies, O. 1959. An Improved Procedure for Starch-Gel Electrophoresis: Further Variations in the Serum Proteins of Normal Individuals. *Biochemistry Journal* 71: 585.

Starikovskaya, Y. B., R. I. Sukernik, T. G. Schurr, A. Kogelnik, and D. G. Wallace. 1998. MtDNA Diversity in Chukchi and Siberian Eskimos: Implications for the Genetic History of Ancient Beringia and the Peopling of the New World. *American Journal of Human Genetics* 63: 1473–1491.

Stephens, M., N. J. Smith, and P. Donnelly. 2001. A New Statistical Method for Haplotype Reconstruction from Population Data. *American Journal of Human Genetics* 68(4): 978.

Tamm, E., T. Kivisild, D. Campbell, et al. 2007. Beringian Standstill and Spread of Native American Founders. *PLOS One* 2(9): e829.

Taylor, C. 2012. *The Black Carib Wars: Freedom, Survival and the Making of the Garifuna*. Oxford: Signal Books.

The 1000 Genomes Project Consortium. 2015. A Global Reference for Human Genetic Variation *Nature* 526: 68–74.

Tilford, C. A., T. Kuroda-Kawaguchi, H. Skaletsky, et al. 2001. A Physical Map of Y Chromosome. *Nature* 409: 943–945.

Torroni, A., R. I. Sukernik, T. G. Schurr, Y. B. Starikovskaya, M. F. Cabell, M. H. Crawford, A. G. Comuzzie, and D. C. Wallace. 1993. mtDNA Variation of Aboriginal Siberians Reveals Distinct Genetic Affinities with Native Americans. *American Journal of Human Genetics* 53(3): 563–590.

Turner, J. H., M. H. Crawford, and W. C. Leyshon. 1975. Phenotypic karyotypic localization of the human Rh-locus on chromosome 1. *Journal of Heredity* 66: 97–99.

Turner, J. H., M. H. Crawford, and W. S. Leyshon. 1976. Localization of the Rh-Trove Point in a Tlaxcaltecan Family by Serological-Banded Karyotype Correlation Studies. In *The Tlaxcaltecans: Prehistory, Demography, Morphology, and Genetics*, vol. 7, pp. 155–160. Lawrence: University of Kansas Publications in Anthropology.

Turner, K. R. 1976. Computer Simulation of Transplanted Tlaxcaltecan Populations. In *The Tlaxcaltecans: Prehistory, Demography, Morphology and Genetics*, vol. 7, pp. 48–58. Lawrence: University of Kansas Publications in Anthropology.

Turner, T. R. 2005. *Biological Anthropology and Ethics: From Repatriation to Genetic Identity*. Albany: State University of New York Press.

Turner, T. R., J. K. Wagner, and G. S. Cabana. 2018. Ethics in Biological Anthropology. *American Journal of Physical Anthropology* 165: 939–951.

Uttley, M. 1991. *Relationship of Measures of Biological Age to Survivorship among Mennonites*. PhD dissertation, University of Kansas.

Uttley, M., and M. H. Crawford. 1994. The Efficacy of Composite Biological Age Score to Predict Survivorship among Kansas and Nebraska Mennonites. *Human Biology* 66: 121–144.

Vajda, E. 2010. A Siberian Link with Na-Dene Languages. *Anthropological Papers of the University of Alaska* 5: 33–99.

Voevodin, A., E. Samilchuk, H. Schatzl, E. Boeri, and G. Franchini. 1996. Interspecies Transmission of Macaque Simian T-cell Leukemia/Lymphoma Virus Type 1 in Baboons Resulted in an Outbreak of Malignant Lymphoma. *Journal of Virology* March 70(3): 1633–1639.

Voight, B. F., S. Kudaravalli, X. Wen, and J. K. Pritchard. 2006. A Map of Recent Positive Selection in the Human Genome. *PLOS Biology* 4(3): e72.

Young, K. 2009. *The Basques in the Genetic Landscape of Europe*. Ph.D. dissertation, University of Kansas.

Young, K. L., E. J. Devor, and M. H. Crawford. 2012. Demic Expansion or Cultural Diffusion: Migration and Basque Origin. In *Causes and Consequences of Human Migration*, pp. 224–249. New York: Cambridge University Press.

Young, K. L., G. Sun, R. Deka, and M. H. Crawford. 2011a. Autosomal Short Tandem Repeat Genetic Variation of the Basques in Spain. *Croatian Medical Journal* 52(3): 372–383.

Young, K. L., G. Sun, R. Deka, and M. H. Crawford. 2011b. Paternal Genetic History of the Basque Population of Spain. *Human Biology* 83(4): 455–475.

West, D., D. H. O'Rourke, and M. H. Crawford. 2010. Introduction: Origins and Settlement of the Indigenous Populations of the Aleutian Archipelago. *Human Biology* 82(5–6): 481–486.

West, D., A. Savinitsky, and M. H. Crawford. 2007. Aleutian Islands: Archaeology, Molecular Genetics and Ecology. *Transactions of the Royal Society of Edinburgh: Earth and the Environmental Sciences* 98: 47–57.

Winterhalder, B., and E. A. Smith. 1981. Optimal Foraging Strategies and Hunter-Gatherer Research in Anthropology. In *Hunter-Gatherer Foraging Strategies: Ethnographic and Archeological Analyses*, pp. 1–35. Chicago: University of Chicago Press.

Zlojutro, M. 2008. *Mitochondrial DNA and Y-chromosome Variation of Eastern Aleut Populations: Implications for the Genetic Structure and Peopling of the Aleutian Archipelago*. Ph.D. dissertation, University of Kansas.

Zlojutro, M., R. Roy, J. Palikij, and M. H. Crawford. 2006. Autosomal STR Variation in a Basque Population: Vizcaya Province. *Human Biology* 78(5): 599–618.

Zlojutro, M., R. Rubicz, and M. H. Crawford. 2009. Mitochondrial DNA and Y-Chromosome Variation in Five Eastern Aleut Communities: Evidence for Genetic Substructure in the Aleut Population. *Annals of Human Biology* 36(5): 511–526.

Zlojutro, M., R. Rubicz, E. J. Devor, V. A. Spitsyn, S. V. Makarov, K. Wilson, and M. H. Crawford. 2006. Genetic Structure of the Aleuts and Circumpolar Populations Based on Mitochondrial DNA Sequences: A Synthesis. *American Journal of Physical Anthropology* 129(3): 446–464.

Index

For the benefit of digital users, indexed terms that span two pages (e.g., 52–53) may, on occasion, appear on only one of those pages.